Management-intensive Grazing

The grassroots of grass farming

Management-intensive Grazing

The grassroots of grass farming

by

Jim Gerrish

with a foreword by Allan Nation

A division of Mississippi Valley Publishing, Corp.
Ridgeland, Mississippi

First printing February 1, 2004
Second printing May 2004
Third printing May 2005
Fourth printing May 2006
Fifth printing March 2007
Sixth printing January 2008
Seventh printing January 2009
Eighth printing March 2010
Ninth printing February 2011
Tenth printing August 2012
Eleventh printing April 2014
Twelfth printing June 2015
Thirteenth printing March 2017
Fourteenth printing May 2017
Fiftheenth printing October 2018

Library of Congress Cataloging-in-Publication Data

Gerrish, Jim, 1956-
Management-intensive grazing : the grassroots of grass farming / by Jim Gerrish ; with a foreword by Allan Nation.
p. cm.
ISBN: 0-9721597-0-3
1. Grazing--Management. 2. Pastures--Management. 3. Range manage ment. I. Title.

SF85.G385 2004
633.2'02--dc22

2003067588

Cover Design by Steve Erickson, Ridgeland, Mississippi

Manufacturered in the United States.
This book is printed on recycled paper.

Table of Contents

To my Father, William Jackson Gerrish,
for allowing me the privilege of growing up on a farm
and teaching me the value of work.

Acknowledgments

No one gets anywhere in life on their own. This book would not have been written if I had not encountered so many individuals on life's pathway who provided assistance and impetus in my life and career. My first thanks go to all those people who said, "That won't work here." Or, "That is too far out of the mainstream." Or, "That's crazy." Thank you for providing the impetus to do something different. Having a challenge is always the first step toward having opportunity. My thanks to Mr. Merten Roestel, rancher from the state of Entre Rios, Argentina, who in the summers of 1977-8 stayed on our family farm and first introduced me to concepts of controlled grazing.

My professional career at the University of Missouri Forage Systems Research Center would never have been as successful without the outstanding support staff stationed there. Special thanks to Dennis Jacobs, Jim Fitzgerald, Floyd Jefferson, Marvin Daniels, and Nadine Henry for doing all that you did for me for so many years. My thanks also to all the students and part time employees who helped collect all that data over the years.

My thanks to Roger Mitchell, Dean of the College of Agriculture, Food, and Natural Resources, who caught the vision of grassland agriculture and allowed and encouraged me to pursue the opportunities therein. My professional colleagues are too many to name. Thanks to you all, but especially to Drs. Ron Morrow, Fred Martz, Jerry Nelson, Craig Roberts, Kevin Moore, and Duane Dailey. My thanks to all the producers I have worked with over the years. The learning was probably always greater on my side. Special thanks to the Green Hills Farm Project for all your ideas and suggestions. Again there are too many individuals to name them all, but special thanks David and Alice, Dennis and Becky, Eric and Hope, Allan and Tauna, and Jim T.

My final thanks go to those who daily shared my successes and endured my failures, but were always there with me. To my wife, Dawn, thanks for always being there. To my children, Chelsea, Ian, Galen, and Cavan, thanks for the help and the joy you have brought. I hope it was fun and you learned a little something along the way.

Foreword

Jim Gerrish and I were having lunch together at the Iowa Sheep Producers conference in 1992, and were talking about semantics. I recounted to him that I had tried to explain to a rancher that our magazine, *The Stockman Grass Farmer*, editorially specialized in "Intensive Grazing Management" as the practice was called at that time.

"Whal, I need a subscription to that because that's exactly how I graze," I mimicked the rancher.

"My cows keep the grass gnawed off right down to the nub all the time. Yup, intensive grazing's exactly what I do."

I told Jim I was appalled that the phrase we were using to describe good grazing management could easily be interpreted as the exact opposite. Jim agreed and sat quietly for a moment thinking.

Finally he said, "You know, I think the words are in the wrong order. What we are really talking about is intensive management. To be more accurate, we really should call it 'Management-intensive Grazing.'"

This was a head-slap moment for me. I loved it!

He was exactly right. The emphasis should be on the word management.

The next month in my column in *The Stockman Grass Farmer* I replayed that story to our readership. I told them that henceforth our method of pasture management would always be called Management-intensive Grazing in *The Stockman Grass Farmer.*

Now, a lot of people didn't like that the acronym for this was the same as a Soviet fighter plane. Others insisted that the word "rotational" be included and some universities modified it to Management-intensive Rotational Grazing, and others shortened this to Intensive Rotational Grazing. However, over the years Management-intensive Grazing has largely won out as the phrase of choice.

I think it only proper that THE book on Management-intensive Grazing be written by the man who first coined the phrase. Jim is particularly well-suited to this task as he is that rare individual who not only was deeply involved in the world of pasture research but also practiced what he preached on his own personal ranch. This made him more credible with real ranchers than most pure researchers.

While in Missouri he was instrumental in starting a multiple-day school for ranchers designed to teach them the core basics of MiG. Many of the people who attended this school have gone on to become recognized MiG teachers in their own right. Missouri rancher, Greg Judy, recounts in his excellent book *No Risk Ranching* how attending Jim's school allowed him to move his small ranch from near bankruptcy to a large thriving ranch in just a few years.

At the *Stockman Grass Farmer*, Jim has had a popular column for years. He is also a favorite speaker and presenter at our conferences and schools. His self-deprecating humor and wry wit have always helped make the medicine go down more easily. I think you will find the same thing true of this book.

Allan Nation
Editor, *The Stockman Grass Farmer*

Introduction to Management-intensive Grazing

1 Management-intensive Grazing Starts with a Lifestyle Goal

Our production systems should enhance the land, not just sustain it.

How would you explain Management-intensive Grazing (MiG) to a visitor from Mars? That was the question Allan Nation asked me, and is the question I am going to try to address in this book. We are encountering a lot of *Homo sapiens* from planet earth who are not much more familiar with the concept than a Martian visitor. This book targets those beginner graziers who are looking for very basic information. That information makes up the grassroots of grazing. While targeting newbies, I also hope that it makes a good refresher for experienced hands who may sometimes forget the fundamentals.

It has been almost ten years since Allan and I were visiting about what we were really trying to accomplish in the grazing community and we started using the term Management-intensive Grazing. A lot of different people have a lot of different ideas about what the term means. A published definition of MiG in the "Journal of Dairy Science" reads: "Management-intensive Grazing is a flexible approach to rotational grazing management whereby animal nutrient demand through the grazing season is balanced with forage supply, and available forage is allocated based on animal requirements."

Just from this fairly simple definition it is apparent that an understanding of animal nutritive requirements, forage supply, seasonality of grazing, and mechanics of rotational grazing must all be incorporated into the management scheme. This biological definition does not consider the economic ramifications of grazing management decisions or the potential environmental impact of MiG. So where do we begin in explaining MiG to our extraterrestrial visitor? Since ET's first question might be, "Why do you do this?" Maybe we should start there.

MiG is a goal-driven approach to grassland management and utilization. While some folks might think the goal of MiG is to have the most miles of fence and number of watering points in the county, that is not the objective at all. As we look at using grassland resources, I think the first goal is one of lifestyle. How do you want to live? The second goal is financial. What is it going to cost to live the way you want? The third consideration is environmental. What do you want your resource base to look like? The final consideration is selecting the production system that allows you to accomplish the first three objectives.

We sometimes hear "real farmers" talk about "lifestyle farmers" in derogatory terms. While lifestyle to some may mean nothing more than a big hat, boots, and a belt buckle, it encompasses much more. It is choices about where you live and work, who you work for, what you eat, when you spend time with your family, and a host of other factors.

Making the choice of living in north Missouri or Jackson Hole, Wyoming has a great deal of impact on how much money you need to earn from the grazing operation. Whether you vacation in Branson or Brisbane dictates how much money you need. All those lifestyle choices affect the financial return required from the farm or ranch. One person's choices may require the typical income from a beef cow-calf operation selling commodity beef while someone else's choices require selling $25,000 elk bulls and high-priced velvet. I adhere to the First Nation philosophy that the land is a loan from our children, not a gift from our fathers. It becomes important that we accomplish our lifestyle and financial goals within the bounds of what our natural resources can sustain. More importantly, our production systems should enhance the land, not just sustain it.

The unfortunate part for many producers is that they start with the production system in place and try to build their goals around an existing white elephant. One great advantage that most beginning graziers have is that they are not tied to an existing set of circumstances. So, think about what the goals of your grazing operation really are and then start planning how you are going to get there. Over the next several chapters, I hope to help you better understand the resources and tools you have available to accomplish those goals.

The Basics:

- MiG is a goal-driven approach to grassland management and utilization.
- The first goal is one of lifestyle.
- The second goal is financial.
- The third consideration is environmental.
- The final consideration is selecting the production system to fit your goals.

Think About Your Farm or Ranch:

- What are your lifestyle goals?
- What are your income goals? Short term/long term?
- What environmental concerns do you face?
- What production system will allow you to accomplish your goals?
- How are you going to get there?

2 Understanding and Building Your Solar Panel

Only green growing leaves
capture solar energy and make cattle feed.

Only green growing leaves capture solar energy and make cattle feed. Bare dirt doesn't do it. Brown dead plants don't do it. This is the first cardinal rule of pasture-based agriculture.

Anyone who has been a reader of the *Stockman Grass Farmer* magazine for more than one issue has read that graziers are solar farmers first and foremost. To be a solar farmer you need to think about building your solar panel.

I like to say that when you buy an acre of land you get 43,560 square feet of solar panel. When you start thinking about your farm in these terms, the importance of having every acre covered with green, growing grass becomes apparent. If you think in terms of square inches and square feet of grass, your focus on growing green leaves becomes an obsession.

While some people already have pasture growing, others may be looking at sowing pasture on bare ground and starting from scratch. In either case, the same basic concepts apply. How you get to where you want to be depends on where you start. Species adaptation is a good starting point. Specific plant species grow in particular climates and landscape positions in response to many years of environmental adaptation and selection. The best choice for pasture in your area is what is growing there naturally or plants from similar climates and environments. While many forage varieties are touted as "superior" or "amazing," most aren't universally adapted to all environments. Many of these "fantastic forages" work well as highly productive, short term crops. If your operation is geared toward reseeding pastures on an ongoing basis, this can work for you. If you are looking for long-term perennial pastures, you should stick with plant species from your own environment.

Understanding plant maturity processes and regrowth following grazing is the next step toward building your "superior" solar panel. In temperate climates there are three basic plant types used for grazing: grasses, legumes, and forbs. Each has different, yet similar, patterns of growth and maturity. In the annual growth cycle of most plants, the end point is seed production. For some plants this is a single cycle while for others multiple cycles occur each year. During each growth cycle, leaves are the most photosynthetically efficient midway through the growth cycle. Early in the growth cycle there may not be enough leaves to capture a high percentage of incoming solar energy. The solar panel is inefficient. Late in the growth cycle there are so many leaves that they shade one another, and lower leaves use more energy for maintenance than they can produce. The solar panel again becomes inefficient. The goal of grazing management is to keep the panel in that in-between stage of solar efficiency. This is where green leaves do their thing.

Thinking about the efficiency of the solar panel brings up another concept. Leaf area index (LAI) is the term used to describe the relative amount of leaf-to-ground area. For instance an LAI of 1 means there is one square foot of leaf surface above one square foot of ground, and so on. For most grass pastures, solar energy capture is optimized at an LAI around 3 to 5. While LAI may increase above these levels, forage yield will increase very little. Once a canopy is intercepting 95% of the incoming solar energy there is little room for improvement.

While we have just considered grasses to this point, many intensively managed pastures also contain legumes or forbs. Forbs is the general term used to describe broad-leafed plants: What a lot of folks would call weeds. Legumes fall into the broad classification of forbs but are generally considered a separate group due to their unique characteristics. Grasses typically have more vertical leaf orientation while forbs and legumes have more horizontal orientation. When the two types are mixed together and the leaves begin to form inter-layers, the efficiency of the solar panel increases. This physical aspect of canopy architecture is one reason why serious graziers talk so much about diversity. Diversity is good for a number of reasons, but building the better solar panel is a very important one. We will talk a lot more about diversity later.

The take home message for this chapter is that green growing leaves are the key to building an efficient solar panel. Adapted species, immature leaves, and species diversity are concepts to build upon.

The Basics:

- When you buy an acre of land, you buy 43,560 square feet of solar panel.
- The best choice for pasture in your area is what is growing there naturally or plants from similar climates and environments.
- In temperate climates there are three basic plant types used for grazing: grasses, legumes, and forbs.
- During each growth cycle, leaves are the most photosynthetically efficient midway through the growth cycle.
- The goal of grazing management is to keep the solar panel in that in-between stage of solar efficiency, not too early, not too late.
- Leaf area index (LAI) is the term used to describe the relative amount of leaf-to-ground area.

- Grasses typically have more vertical leaf orientation while forbs and legumes have more horizontal orientation.
- Diversity of plants for grazing is good for a number of reasons, but building the better solar panel is a very important one.

Think About Your Farm or Ranch:

- What native grasses, forbs and legumes are already thriving on your ranch?
- What other plant species would grow in your environment?
- Which plants can you use from similar climates or environments?

3 The Role of Animals in Management-intensive Grazing

Understanding how animals interact with their natural and manmade surroundings is critical to maintaining healthy, productive animals.

Many prominent graziers have worked for the last couple of decades to get live stock producers to quit thinking in terms of being cattlemen, dairymen, or sheepmen and start thinking of themselves as grass farmers. Of course, another whole segment of the industry has been trying to get all of us to think in terms of cattle-persons, not cattlemen, and so forth.

While the grass farmer concept is critical to shifting emphasis to grass management as the primary resource in pasture-based production systems, it can be carried too far and the role of the animal in the system slighted. The really important concept is to keep all things in balance and not over emphasize any component to the detriment of others.

Once the solar energy has been converted into grazeable forage, the animal needs to step forward and carry out the next step in the process: converting solar energy to milk, meat, or fiber. In this chapter we will begin discussing the grazing animal and its needs.

Livestock needs can be broadly broken down into three categories: nutritional, environmental, and social needs. Some experts would add a fourth category and that is health care needs, which would include vaccinations, parasite control, and therapeutic treatments. I believe that the three primary needs interact to determine the overall well-being and health status of the animal. The degree to which the typical "health program" is required on a particular farm or ranch is dependent upon how well the three primary needs areas are being met.

While ruminants are very different physiologically from monogastrics, or single stomached animals, their basic nutritional needs are very similar to humans. In our own diets we are concerned about protein, energy, minerals, vitamins, and water. The same broad categories apply to ruminants. The unique characteristic of ruminants is how they are able to use sources of nutrients that are unavailable to monogastrics.

Ruminants have four stomach compartments, largest of which is the rumen. The rumen is a fermentation vat which ranges in size from a few gallons in smaller breeds of sheep and goats to thirty or more gallons in large cattle. Living inside this fermentation vat are billions of microbes including bacteria, fungi, and protozoans. It is these little critters which are able to take apart complex plant fibers and extract the nutrients needed by grazing livestock. Maintaining a healthy environment for the microbes is an important part of the grazier's job.

Keeping the rumen healthy and functioning includes maintaining stable pH or stomach acid level, carbon to nitrogen (C:N) ratio, and providing required minerals and vitamins in appropriate amounts. The nice thing about this job is that most mixed grass-legume pastures maintained in a vegetative condition growing on soil of moderate fertility will provide just the right combination of nutrients to keep things ticking along nicely. Thus the grazier's job can be very easy. Problems arise when the pasture is not properly diverse; when soil nutrients are inadequate, in excess, or otherwise out of balance; or when some other source of nutrients besides pasture is being provided to the animal. Future chapters will deal with meeting these challenges on an individual basis.

Some basic nutritional relationships to keep in mind follow. Vegetative forage is usually high in protein and energy while mature forage is low in both. Minerals may be higher in total amount in mature forage, but they are usually more readily available to the animal in less mature forage. As cool-season forages mature, energy tends to be the most limiting nutrient while protein tends to be the most limiting in warm-season forages as they mature. All of these factors are considerations in making grazing management and supplementation decisions. More about these relationships later.

The environmental needs of livestock deal with their adaptation to prevailing environmental conditions, both natural and manmade. Selection of animals with appropriate tolerance levels of heat and cold, high or low humidity, insects and pests, and other factors present in your environment is essential to long term sustainability in pasture-based enterprises. As we all know, humankind has altered the environment for livestock in both positive and negative ways. The presence or absence of shade, availability and quality of water, location of paddocks within the changing landscape are all factors that come readily to mind. Understanding how animals interact with their natural and manmade surroundings is critical to maintaining healthy, productive animals.

Understanding animal interaction with their environment leads right into the third category of animal needs which is their social needs. The *Stockman Grass Farmer* magazine has done an outstanding job of highlighting the work of such individuals as Bud Williams, Burt Smith, and Temple Grandin to bring us all a better understanding of the social needs of our livestock. The impact of social stress on the productivity and performance of livestock was largely ignored by the livestock industry until recent years.

The impact of rough handling on everything from cow disposition to meat quality is becoming better understood. Even what many of us considered to be fairly gentle handling over the years has been shown to affect animal performance for much longer periods of time than we ever dreamed. As this publication will undoubtedly continue to publish many articles on low-stress livestock handling, I will leave that area to persons more expert than me. Next we will begin to explore the relationship between the pasture and the grazing animal and how our management decisions impact that relationship.

The Basics:

- Forget about the species you run. Think of yourself as a grass farmer.
- Grass management is the primary resource in pasture-based production systems.
- Keep all things in balance.
- Livestock needs can be broadly broken down into three categories: nutritional, environmental, and social needs.
- Maintaining a healthy environment for the rumen's microbes is an important part of the grazier's job.
- Most mixed grass-legume pastures maintained in a vegetative condition growing on soil of moderate fertility will provide just the right combination of nutrients that ruminants need.
- Vegetative forage is usually high in protein and energy while mature forage is low in both.
- Minerals may be higher in total amount in mature forage, but they are usually more readily available to the animal in less mature forage.
- As cool-season forages mature, energy tends to be the most limiting nutrient. In contrast, protein tends to be the most limiting nutrient in warm-season forages as they mature.

Think About Your Farm or Ranch:

- What are you doing to meet your animals' nutritional, environmental and social needs?
- What species and breed of animal are most suitable for your climate and environment?

4 Why Should You Divide Pastures Into Paddocks?

With MiG we seek to intensify management, not necessarily grazing.

One of the hallmarks of MiG operations is that big pastures are divided into little pastures or paddocks. We see paddocks of all different sizes and shapes. Some farms have fewer than ten while other farms have a hundred or more. With so many examples available and so much variability, we need to ask ourselves just why are they dividing these pastures?

The simple answer is for management control.

Always remember that with MiG we seek to intensify management, not necessarily grazing. Why there are so many different levels of subdivision occurring on neighboring farms and ranches comes down to the basic issue of how much management control do you want or need to have?

There are several factors affecting how much control may be needed to help you accomplish your goals.

Let's first look at what subdividing pastures allows you to accomplish.

When you have just one pasture, livestock are free to graze wherever they want, whenever they want, for as long as they want. Pasture management is largely being left to the animal's good judgement. While a cow may be very good at being a cow, she is not a reliable manager for long-term pasture health within a single fenced enclosure.

Before pioneer days, the buffalo were pretty good pasture managers because they had the opportunity to leave where they were and go somewhere else when the pasture had been grazed sufficiently. Once an area of range had been grazed and fouled with manure, there was little reason for the buffalo herd to stay there and they would move on to greener pastures.

Today, buffalo on fenced ranches are no better pasture managers than the common cow. Because the cow or buffalo no longer has the option to leave the pasture when they need to, we must do something to create a similar response to animals leaving the range where they have been. Moving the animals to another paddock within the pasture provides that opportunity.

The most common argument for rotation grazing is allowing a pasture to rest. When we have multiple paddocks within the pasture, we create the opportunity for the animals to go somewhere else. When the animals go somewhere else, the plants have the opportunity to grow new leaves. As new leaves grow, more photosynthesis takes place and the overall energy status of the plant improves. Root growth is renewed and the plant becomes more robust.

Some plants are more grazing tolerant than others and require little rest. These are species that have a lot of leaves close to the ground. Other species require more rest because they have fewer leaves close to the ground. These are the sorts of things that determine how long a pasture needs to rest between grazings to stay healthy.

Providing appropriate rest periods allows the less grazing tolerant species to stay productive in a mixed pasture. From the rest perspective, how many paddocks needed is determined by the number of days spent grazing each paddock and how many days of rest are required. For a simple pasture mixture, that may be as few as four to eight paddocks.

If we only need eight paddocks, why do we see some graziers with 20, 40, or 100?

There are other goals besides just providing rest. Controlling forage quality, regulating grazing intake, managing grazing efficiency or utilization rate, and manure distribution are all important considerations for some operations. It all comes back to what your goals are and what you are trying to accomplish through MiG.

For a commercial cow-calf operation having a goal of improving conception rate and weaning weights through better pasture quality, by just subdividing to eight paddocks can allow them to consistently keep red clover in the pasture. Having a legume in a grass pasture can add 30 to 100 lb weaning weight and typically increases conception rate.

If the same operation decides they want to graze all year around and feed no hay, then the level of subdivision required increases substantially. There needs to be a number of paddocks rested in late summer and fall to allow stockpiling winter forage, and the length of grazing period in the winter must be shortened. The year-around grazing scenario may require 20 or more paddocks.

What if they decide to keep their calves and background them? Forage quality becomes a more important issue than it was when they were just raising cows and calves. Forage quality can be better controlled with a higher level of subdivision and frequent cattle moves. Once again, the level of subdivision may need to increase.

Looking at dairy grazing, it has been repeatedly demonstrated that more frequent rotation to fresh pasture increases milk production. Dairy farms tend to use the highest level of paddock subdivision compared to other enterprises. But you have been to pasture-based dairy farms that don't look like they have very much fence. There are different ways of achieving a high level of subdivision without a lot of permanent fence. Movable electric fencing provides great flexibility in managing paddock size and numbers.

Management flexibility is a fundamental concept of MiG. The greater the level of subdivision opportunity, the greater the management flexibility, but too much permanent subdivision can easily become restrictive. The key to successful grazing management, is knowing what you are trying to accomplish and maintaining adequate flexibility to achieve your goals.

The Basics:

- The main reason for pasture subdivision is management control.
- When we have multiple paddocks within the pasture we can allow the pasture to rest.
- Providing appropriate rest periods allows the less grazing tolerant species to stay productive in a mixed pasture.
- Controlling forage quality, regulating grazing intake, managing grazing efficiency or utilization rate, and manure distribution are all important considerations for some operations regarding the number of paddock subdivisions.

Think About Your Farm or Ranch:

- How much management control do you want or need to have?
- What are your goals, and what are you are trying to accomplish through MiG?

From the Ground Up — Soils

5 Soil — Balancing the Three-Legged Stool

Soil is a three-legged stool consisting of the physical structure, biological health, and chemical nature of the soil.

It has been said by men far wiser than I am that life as we know it on this planet is tied to the thin layer of top soil which covers most of the continental landmass. When viewed in proportion to the whole mass of the earth, soil is a microscopically thin layer.

Soil in some locales is measured in inches, in contrast to the 8,000 miles of the earth's diameter. I think there is no nobler task for a person than to take up the responsibility for keeping this treasure intact and healthy.

I like to think of the soil as a three-legged milking stool that your farm or ranch sets atop. The three legs are the physical structure, biological health, and chemical nature of the soil. Unfortunately, for many years too many people thought only in terms of the chemical nature of the soil.

While soil testing is a very important management tool, a lot of people have had the idea that if you took a soil test and applied fertilizer as recommended, you had taken good care of the soil. That ignores the other two legs of the stool.

When one leg of the stool becomes longer or shorter than the other two, the pasture goes out of balance and the farm starts sliding from the top of the stool. Something has to be done to bring it back into balance.

Too often the farm is brought back into balance by shortening the other two legs, rather than raising them. The shortening of the other legs is often not the result of deliberate management but, rather, by the absence of management which allows the other legs to sink to a lower level. We can make management choices which affect each of the three legs.

When we make a decision about where we are going to live, we have already made some basic choices regarding the physical nature of the soil we have to work with. Soil depth and soil texture, whether we have sand or clay, is dictated to a large degree by geographical or topographical locations. However, there are other factors that we affect within the bounds of soil depth and texture.

Soil compaction is a concern for many graziers and it is affected by grazing management. Three factors determine degree of soil compaction: soil type, soil moisture, and force applied to the soil.

Grazing pastures on certain soil types in wet conditions can result in severe soil compaction. Many graziers have a fear of what happens when they graze pastures during droughts and wonder what the long term impact on the soil is.

Unless the soil is completely stripped and it blows away, the damage to the soil by grazing when it is dry is very minimal. Pastures recover remarkably quickly when rain finally comes following a drought.

The same soil and pasture may not recover for a couple of years if grazed when too wet.

Grazing when too wet collapses pore structure in the soil leading to reduction of both water infiltration and holding capacity as well as oxygen holding capacity. Root growth is also reduced due to greater physical impediment to growth.

Organic matter is another essential part of soil structure. While we tend to think of organic matter (OM) as the biological life of the soil, it is also critical to the physical structure of the soil. Much of the infiltration capacity and water holding ability of a soil is related to OM content.

Grazing that keeps pasture vegetation short results in loss of OM from the soil while grazing management that allows vigorous top and root growth generates additional OM.

Tillage tends to oxidize organic matter which is one of the reasons many graziers hate the idea of tilling a well established pasture.

Tillage also destroys the root channels and worm burrowings that have developed over the years, both of which help water infiltration and aeration.

Organic matter also ties closely to the chemical nature of the soil as it is a large pool of soil nitrogen (N) and other essential minerals. Negatively charged organic matter provides additional binding sites for minerals in the soil.

Soil texture also determines the nutrient holding capacity of the soil with clay particles having many more binding sites than sand.

The amount of minerals in the soil also affects biological activity in the soil, as does water supply, which is affected by both soil texture and organic matter.

Hopefully, you now see why I view soil as that three-legged stool. Physical, biological, and chemical factors are all tied together. Shorting one factor affects the stability of the other legs, but by the same token, management that enhances one factor is likely to also enhance the other two.

Our management can build the soil up to higher and higher levels as long as we remember that there is more to soil management than just taking a soil test and applying fertilizer.

The Basics:

- Soil depth and soil texture, whether we have sand or clay, is dictated to a large degree by geographical or topographical locations.
- Three factors determine degree of soil compaction: soil type, soil moisture, and force applied to the soil.

- Infiltration capacity and water holding ability of a soil is related to OM content.
- Grazing that keeps pasture vegetation short results in loss of OM from the soil while grazing management that allows vigorous top and root growth generates additional OM.
- Tillage tends to oxidize OM, destroys the root channels and worm burrowings that have developed over the years, and thus inhibits water infiltration and aeration.
- The amount of minerals in the soil also affects biological activity in the soil, as does water supply, which is affected by both soil texture and organic matter.

Think About Your Farm or Ranch:

- Assess the soil of your farm. Is the physical structure, biological health and chemical nature of the soil balanced?
- What can you do to raise the shorter legs of your soil?

6 Mud Balls, Sand Castles, and Soil Compaction

Several factors interact to determine whether or not soil compaction is going to be a problem in a pasture.

About ten years ago as I was moving polytape fences in one of our grazing studies I noticed the step-in posts were a lot harder to put in the ground in some pastures compared to others. The study compared continuous grazing with weekly or daily rotational grazing on big bluestem.

In 1991, the first year of the study, spring and early summer moisture were near normal to slightly above normal while mid to late summer moisture was quite low. What I observed from stepping in posts was that the ground was much harder in the continuously grazed pastures compared to rotationally grazed pastures.

Well, that made some sense because we had significantly more top growth on the daily rotation pastures compared to continuously grazed pastures and that should have translated to more root growth.

But, on the other hand, more top growth would also mean a higher transpiration rate and more moisture being pulled out of the soil in the rotationally grazed pastures. That first year we made no measurements of soil compaction, but I began to think about it a little more.

The following year we had a dry spring followed by 11 inches of rain in July and then a return to below normal rainfall. By the time mid-August rolled around, you could hardly get a post into the ground in the daily rotation pasture.

We took one set of soil bulk density measurements in late August by extracting soil cores from each of the grazing treatments and also under the fence lines between each pasture where no cattle could have stepped. When our silt loam soils are in a healthy, mellow state, the bulk density will be between 1.05 and 1.25 grams per cubic centimeter (g/c-cm).

The samples we took from the pastures were all around 1.35 to 1.4 g/c-cm. while the samples from under the fence line were all about 1.15. The next year we decided we should take bulk density samples throughout the grazing season and try to assess what was going on regarding soil compaction in different grazing systems.

Unfortunately, the next year was 1993, which all Midwesterners remember as the year of the Great Flood. It began raining in the middle of May and did not quit until September. We had 32 inches of rain in June and July and we took no soil density measurements because it is really challenging to keep gooey mud in the soil cores.

What this little experience did show us was that several factors interact to determine whether or not soil compaction is going to be a problem in a pasture. There are three basic factors that determine the likelihood of soil compaction developing: soil type, soil moisture, and physical force applied to the soil.

Some soils are going to just be a lot more prone to compaction than others. If you are going to have a mud ball fight with your brothers, do you want to be on the clay knob or down in a sandy wash?

Clay can be compacted much more easily than sand because of the fineness of the soil particles. All other things being equal, the order of susceptibility to compaction is clay>clay loam>silty clay>silt>silt loam>sandy loam>sand. Of course a real soil scientist would have had 32 other soil textures in that array but this gives a general idea of what is going on. Another important textural factor affecting compaction potential is soil organic matter. Organic matter is like a sponge and provides resilience to the soil. Within any class of soil, a higher level of organic matter reduces propensity for compaction and also allows the soil to recover more quickly if compaction does occur. Even above-ground organic material in the pasture helps minimize compaction.

Leaving a taller post-grazing residual reduces compaction by providing more physical cushioning for the animals' hooves as well as maintaining a greater amount of root growth potential. The dead plant residue layer on the soil surface also provides cushioning and helps protect the soil.

Soil moisture is the next big factor in determining compaction. Is soil damage more likely to occur during a drought or the rainy season? Little change in soil density occurs during a drought because it is very hard to compress a dry soil until the plants are killed and the roots die. But it is very easy to compress a wet soil.

Why do we have to go to the beach rather than the desert to build sand castles? Because it is awfully hard to make dry sand take on some other form than loose dry sand, but even sand can be compacted with the right amount of moisture.

How are bricks made? Through the combination of clay, water, and heat. If you want to see a lot of compacted soil in a pasture go to anyplace with clay-based soil that typically experiences high spring rainfall followed by hot, dry summers. Sounds a lot like a typical Missouri summer!

The third compaction factor is physical force applied to the soil. In pastures that force takes the form of hooves striking the soil, as well as any mechanized activities taking place on the pasture such as clipping, dragging, fertilizing, motocross, etc. Physical impact from grazing is the accumulated hoof beats times the mass/square-inch of each foot step. How far does a cow typically walk in a day? It depends. What it depends on are opportunity for strolling, forage availability, terrain, and type of livestock. Most livestock like to check out their surroundings on a regular basis. Size of pasture has little bearing on how far an animal will walk in a day until forage becomes limiting.

Research conducted by Mike O'Sullivan in Ireland during the 1980s found cattle in both continuous and rotational grazing systems walked the same distance per day, until forage supply began to decline. Once foraging to meet their daily needs became more challenging, continuously grazed cattle walked up to three times farther per day than their rotationally grazing counterparts.

In this study, rotationally grazing cattle maintained a fairly steady rate of three to five miles per day, while the continuously grazed cattle increased their travel distance up to eight to ten miles per day. Increased travel distance translates directly to increased hoof beats and increased physical force exerted on the soil. Beef cows tend to walk farther every day than do dairy cows.

So with all this in mind, what was going on in our grazing study in the early 1990s?

During the first year when early season moisture was adequate but late season moisture was inadequate, the primary effect that we were seeing was probably due to reduced plant growth in continuously grazed pasture and increased travel distance as forage supply became limited. Thus, we had a classic case of compacted soil in continuously grazed pastures while rotational grazing maintained a more mellow soil.

In the second year when it was dry early and very wet in July, the rotationally grazed pastures became compacted due to restricted plant growth early followed by excessive physical force (stock density of 45,000 lb animal liveweight/acre) on a muddy soil (11 inches of rain in 31 days).

What is the practical application of this knowledge?

On a farm with variable soil types, plan the grazing season to concentrate use on sandy or loamier soils when it is wet and avoid heavy clay soils. Shorten grazing periods during wet periods to reduce total hoof impact on the soil.

Remember, based on O'Sullivan's work, dispersing animals to a larger area may not reduce animal impact, particularly if pasture is short. Grazing management that leaves ample post-grazing residual, maintains soil surface residue cover, and maintains or increases soil organic matter level will help protect your pastures from compaction potential.

After failing to do an adequate job of monitoring soil compaction during the big bluestem grazing study in the early 1990s, we set out to compare soil compaction under continuous and rotational grazing at four different stocking rates during an April to September stocker program over a five-year period from 1996 through 2000. What we found was increased stocking rate resulted in increased soil compaction regardless of grazing method.

Why? More hoof beats on the ground every day at higher stocking rate. Rotational grazing did tend to moderate the effect because of more plant cover and soil residue throughout the grazing season. Each season we saw soil density increase from spring through the grazing season and each winter freezing and thawing send soil density back to its spring level. Some years the cycle was not perfect, some years the following spring density was lower than the previous year. It gives those of us who live in more northerly climates one more thing to be thankful for compared to southern graziers, our soil compaction problem will pretty well take care of itself with winter rest.

The Basics:

- Factors determining the likelihood of soil compaction developing are soil type, soil moisture, and physical force applied to the soil.
- Increasing soil organic matter reduces compaction potential.
- Little change in soil density occurs during a drought because it is hard to compress a dry soil until the plants are killed and the roots die. But it is very easy to compress a wet soil.
- In pastures physical force takes the form of hooves striking

the soil, as well as any mechanized activities taking place on the pasture.

- Increased stocking rate results in increased soil compaction regardless of grazing method.

Think About Your Farm or Ranch:

- What type of soils are in your pasture?
- How does your seasonal rainfall or lack of moisture affect your pastures?
- When should you shorten, concentrate, or lengthen your grazing?
- How can you change your grazing management to protect your pastures from soil compaction?

7 A Walk in the Rain — Managing the Water Cycle

If you allow pastures to grow roots and maintain good ground cover, infiltration will increase, and runoff will decrease.

I had an interesting experience a couple of years ago one morning as I was driving from home to work at the Forage Systems Research Center.

Like many grass farmers and university professors, I don't drive a new vehicle. The particular rig I was in that morning had 146,000+ miles on it and had been having some electronics problems over the last few months. One of the affected components was the fuel gauge which indicated I had a quarter tank of gas. This vehicle usually runs out between 1/16 and 1/8th tank so I watch it pretty closely.

Well this morning it ran out at a quarter tank about a mile and a half from home. It had been fairly dry for several months but we were finally getting some much needed rain. Over an inch had fallen in the first hour already. So I started my trek back towards home in pretty steady rain. Someone who does not enjoy walking as much as I do may have actually been upset by the situation. Anyone who knows me would of course assume that I would not walk along the road to go home but would jump the fence and head cross country as soon as possible and that's exactly what I did.

If I had walked along the road, maybe someone would

have seen me and I would have to explain that I had run out of gas. Then they would likely say, "You know they run just as well on the top half as the bottom half." Of course on my road, the likelihood of anyone coming by was fairly remote.

The pasture I first crossed into belonged to one of my neighbors and was usually grazed pretty heavily for seven to ten months every year. It's one of those pastures you can drive by on the road and see a rabbit at a hundred yards. It was a nice rain and I noted a lot of water running on the surface and the ditches were beginning to flow. Water was beginning to stand in the low spots. Because of the short water situation resulting from a year of drought, the neighbor had a new pond dug in April. That pond had filled with a few of the rains that had come in June.

And then I crossed the fence onto my own farm.

This was on the same soil type. Both areas had been in pasture for many years and the same rain was falling on both sides of the fence. But there was a night and day difference in what was going on between the two pastures. The neighbor's pasture was continuously grazed and mine was MiG.

You know how the text books say if you allow pastures to grow roots and maintain good ground cover, infiltration will increase and runoff decrease? Well, it's true. As soon as I crossed the fence, there was no water moving laterally on the surface and no ponded areas were visible while the rain still poured down. No water was moving in any of the stream channels in the pasture and my pond was still six feet below the spillway.

Previously, we discussed soil compaction in pasture, its causes and how to deal with it. Compaction is the factor which most limits water infiltration because the effect of compaction is collapse of soil pore structure where water flow occurs. Water holding capacity of the soil is inversely related to compaction. All factors that help determine the extent of compaction also help determine infiltration rate and water holding capacity, which is the foundation of a healthy water cycle. To build a healthy water cycle in pasture you need to start by knowing what the soil potential and limitations are and then target grazing management to capitalize on the potential and reduce the limitations. Decisions from what type of forage to plant to timing of grazing can affect the water cycle on specific soil sites. There are some fairly universal rules that can be applied to almost all soil environments and plant communities that will enhance the water cycle.

Rainfall is first intercepted in the upper part of the plant canopy. The greater the percentage of the grazing land that is being rested at any time, the more acres that are covered with a taller, more highly structured canopy. The upper canopy breaks the impact of raindrops and funnels water along leaves, then on to stems, and down toward the plant base or crown where the greatest concentration of roots occurs.

In bunch grass communities, soil organic matter is almost always higher nearer the bunch than away from the bunch, thus infiltration occurs more rapidly for water that reaches the soil surface nearer the plant base. Shorter grazing periods with multiple paddocks ensures that a high percentage of the total plant community surface is being rested and is in a taller state. Leaving a taller post-grazing residual keeps soil exposure minimized and also maintains the physical barrier to raindrop impact and retards overland flow.

Both pasture rest and leaving adequate post-grazing residual also promote more extensive root systems and leave more plant residue on the soil surface. The residue is one more layer to soften the impact of rain and help keep water from leaving the point of impact so rapidly. Proper rest, residual, and residue management are the keys to maintaining a healthy water cycle in your pastures. Particularly here in the Midwest, most droughts are acts of man, not acts of nature.

After that storm passed over my neighbor's pasture and mine, I still had the problem of limited livestock water in the pond, but I was confident my pastures would continue to grow a few weeks longer than my neighbor's when the rain quit.

The Basics:

- Compaction is the factor that most limits water infiltration.
- Rainfall is first intercepted in the upper part of the plant canopy. The greater the percentage of the grazing land being rested at any time, the more acres covered with a taller, more highly structured canopy.
- In bunch grass communities, soil organic matter is almost always higher nearer the bunch than away from the bunch, thus infiltration occurs more rapidly for water which reaches the soil surface nearer the plant base.
- Shorter grazing periods with multiple paddocks ensures a high percentage of the total plant community surface is being rested and is in a taller state.
- Leaving a taller post-grazing residual keeps soil exposure minimized and also maintains the physical barrier to raindrop impact and retards overland flow.
- Both pasture rest and leaving adequate post-grazing residual also promote more extensive root systems and leave more plant residue on the soil surface.
- Proper rest, residual, and residue management are the keys to maintaining a healthy water cycle in your pastures.

Think About Your Farm or Ranch:

- What is the soil potential and limitations of your pastures?
- How can you target grazing management to capitalize on the potential and reduce these limitations?

8 The Importance of Making Pasture Fertility Pay

Pasture fertilization can be a black hole to pour money into or it can be a profitable management tool.

Over the years I have heard many cattlemen say. "It doesn't pay to fertilize pasture."

What makes this statement so interesting is that sometimes it can be quite true while other times it is the farthest thing from the truth. So what makes pasture fertility pay? I think we need to divide fertility into four categories as we consider the question.

The first category is lime and soil pH. Pasture and range soils in the USA range from pH 4 to 12, depending where in the country you are located. Very acid soils and very alkaline soils offer unique challenges to graziers in those regions. The most productive pastures tend to be near neutral soils, all other factors being constant.

In the humid part of the country, where pastures tend to be acidic, liming almost always pays. There are several reasons for this. Most of the soil minerals essential for plants' growth have their highest availabilities in slightly acid to slightly alkali soil. As pH increases and soil becomes more alkali, non-essential minerals and compounds with potential animal toxicity are more likely to increase in plants. With more acid conditions, minerals toxic to the plant tend to increase in availability.

Neutralizing the pH produces a better balance of essential minerals.

Microbial activity in the soil is maximum near neutral pH resulting in more efficient nutrient cycling. Rhizobia bacteria responsible for nitrogen fixation are very sensitive to soil pH and are much less active when pH drops below 5. Lastly, liming provides calcium essential for plant growth and, in magnesium deficient areas, dolomitic lime can also provide magnesium.

I think lime is the first step in rebuilding damaged soils. Over 3/4ths of our farm was in row crop when we bought it in 1986 and was severely eroded. For most of the farm, the pH range was 4.2 to 5.2. On part of the farm, lime is the only soil amendment we have applied. On those areas carrying capacity has steadily increased.

Because of the effects of soil pH on so many other processes, just liming can really make a difference on almost any pasture in the humid part of the country. We started applying two tons per acre every three or four years to reach the required 6 to 7 tons to raise pH to the low 6's. All of the lime haulers in our area now have a three ton minimum spreading rate unless you want to pay a higher price per ton so we now apply three tons. We still have not gotten the entire farm covered but are slowly working at it.

The second fertilizer category is phosphorus (P) and potassium (K), the other two players in the big NPK team. In many parts of the country, P is the second most limiting nutrient for plant growth, while in a few areas it is excessive. All energy transformation in plants is P dependent, which is why it is so important. Like most nutrients, plant response to soil P level is a diminishing returns curve. The lower down on the curve you are, the greater the response to applied P (see figure next page).

If you already have a high P soil, the likelihood of getting an economic response to P fertilization is very slight. If P is limiting in the soil, the usual way to increase its availability is applying either commercial fertilizer or manure. But if you recall from the lime discussion, most nutrients are more avail-

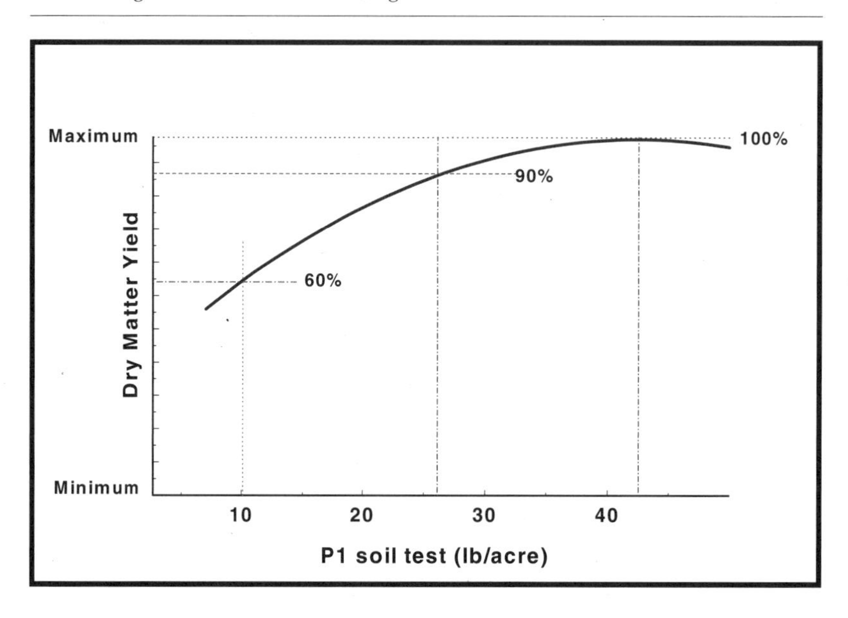

able near neutral pH. Phosphorus is a classic example of this. Raising soil pH from 4.5 to 6.5 can significantly increase available P in the soil. By reverse logic, applying P fertilizer to a low pH soil is throwing money down a hole because of the capacity of acidic soils to bind P into unavailable forms.

Liming and P fertilization go hand in hand in creating productive, high quality pastures.

Most legumes are very sensitive to both pH and P levels. If you plan to switch from N-based grass pastures to grass-legume mixtures, liming and P fertilization are critical management steps. Potassium is helpful for legume persistence and disease resistance. While grass pastures require relatively low K levels, legume pastures need much more. With proper pH, P, and K levels legume-based pastures are just as productive and higher quality than are N-based grass pastures. Records at the Forage Systems Research Center (FSRC) in Linneus, Missouri, show that the annual fertility costs for legume based pastures are typically less than half that of N-based pastures.

Which brings us to the third fertilizer category which is nitrogen (N). Funny thing that most pasture fertilizer talks focus on N and sometimes the presenter will get around to the other nutrients. Yes, N is the most limiting nutrient for plant growth. Yes, N is the least stable nutrient in the soil environment and it must constantly be replenished. But it is also the nutrient that can come from more sources than any others.

I think when producers say that pasture fertilization does not pay they are most likely thinking about the big bill they just got from the fertilizer plant with most of it going to pay for the N. In the South where bermudagrass is the dominant grass and a lot of winter annuals are used, very large quantities of N fertilizer are applied and with proper grazing management can be quite profitable. In some Midwestern legume-based pastures, no N is applied and they can be quite profitable. Somewhere, somehow the pasture has to have an adequate source or sources of N to be productive.

Timing of N application and grazing management have a lot to do with the profitability of N fertilization. In the Midwest a pound of grass has much more economic value in January than in June. Spring applied fertilizer on cool-season grasses just adds to the excess pasture problem in the spring of the year.

The alternative to fertilizer-stimulated grass in June is unfertilized grass which often comes at a lower cost. In contrast, the alternative to N-stimulated stockpiled grass in January is a bale of hay which comes at a substantially higher cost. In our environment, late summer applied N for stockpiled pasture almost always pays while spring-applied N almost never pays. When N prices ran up to 35¢/lb last winter, I checked to see what we could afford to pay for N for stockpiling. To my shock, I found that we could pay over 60¢/lb for N before we reached the breakeven cost of feeding hay.

The final fertilizer category is micronutrients. Besides N, P, and K there are about 15 or so other mineral nutrients that are considered to be essential for plant growth. Not all are required by all plants, but all have been shown to be required by some plants.

Micronutrient availability is very site and soil dependent.

It is nearly impossible to make any kind of blanket recommendation relative to micronutrient application. In the limited amount of work that we have done, we have seen no responses in yield, quality, or persistence to micronutrient applications on pasture.

If there is reason to believe that your pastures may have a deficiency, use a judicious combination of soil and tissue sampling to identify the problem. Apply the needed nutrient to a limited area, monitor the results, and determine whether it was worth the cost.

Pasture fertilization can be a black hole to pour money into or it can be a profitable management tool. To try to make it as profitable as possible, start by knowing what you have to work with through systematic soil testing, determine appropriate goals for your resources and enterprises, make amendments and monitor the results, and then make adjustments as needed.

The Basics:

- Fertility should be divided into four categories: lime and soil pH, phosphorus and potassium, nitrogen, micronutrients.
- The most productive pastures tend to have near neutral soils.
- In the humid part of the country, where pastures tend to be acidic, liming almost always pays.
- Neutralizing the pH produces a better balance of essential minerals.
- Lime is the first step in rebuilding damaged soils.
- All energy transformation in plants is P dependent which is why it is so important.

Think About Your Farm or Ranch:

- Have you done a soil test on your pastures?
- What are the goals for your enterprises and resources?
- Plan to make amendments and monitor the results.

9 Managing the Nutrient Cycling of Grazing Animals

Understanding patterns of consumption and excretion is an important tool for graziers to manage nutrient cycling.

One of the really cool things about pastures and grazing compared to other agricultural systems is that we have the capacity to use the same atoms or molecules over and over.

If we look at fertilizer applied to a corn crop, for example, we see a large part of our investment leaving the farm in the marketed crop. If we apply no fertilizer, the soil is soon depleted of essential nutrients. The worn out cotton fields of the Old South are a classic example of soil exploitation with no return.

In contrast, animals grazing on pasture return over 90% of the nutrients they consume back to the soil through their urine and feces. Fertilizer P and K applied one time can keep working for us for many years, with the right grazing management. Understanding patterns of consumption and excretion is an important tool for graziers to manage nutrient cycling on their farm or ranch.

Not all minerals flow the same pathways within the animal. For example, almost all phosphorus passes through the feces while most potassium flows through the urinary system. This results in different patterns of redistribution within a pasture because livestock do not urinate and defecate simultaneously.

Urine distribution is much more uniform across a pasture than is defecation because animals tend to urinate more frequently while they are actively grazing, while most defecation occurs during or following rest. A mature cow will urinate 15 to 20 times a day while defecating about 10 times daily. The result is nutrients flowing through urine have a better distribution pattern in the pasture and nutrients flowing through feces tend to concentrate at animal rest spots.

When the animal's diet is near a required protein level, excreted N will be about evenly divided between urine and feces. The form of N in urine is almost all immediately available for plant uptake while N in feces is in more complex organic forms and is released more slowly over time.

As the protein level of the diet rises above animal requirement, which is frequently the case in high quality pastures with a strong legume component, the amount of N in the feces remains near constant and all the excess N passes through the urine. The N fertilizer equivalent in a urine spot in high protein pasture can be in excess of 1000 lb-N/acre. And you wondered why those green spots still showed up even when you applied N! With basic understanding of where the minerals flow, we can begin to plan how to manage nutrient cycles in our pastures. As mentioned before, distribution within the pasture is not perfect. Several manageable factors affect manure distribution patterns.

Let's look first at the pattern in a large pasture with uncontrolled grazing. Let's assume we have 300 acres with a single water source from a pond located near the gate into the pasture. The maximum travel distance to water is between 1/2 and 1 mile so the entire herd is likely to come to water together. Because of convenience, salt and mineral supplements are fed near the pond because it is close to the gate. If there is shade near the watering point, about 2/3 of the defecations of the entire herd are likely to occur within a few hundred feet of the shade and water source. Because of the low stock density associated with uncontrolled grazing, excretions are likely to occur on only 2 to 5 % of the pasture area each year. This all makes for a very inefficient nutrient cycle.

Shade, water, and supplemental feeding sites are manure magnets. The fewer water sites and the farther between them, the less efficient the nutrient cycle. Conversely, the more watering sites and the closer they are, the more efficient the cycle becomes. This is largely due to changing grazing distribution patterns but also due to breaking up the herd's social behavior of all going to water together. Greater travel distances to water encourage herd behavior and also cause the herd to stay longer at the watering point. This results in a higher percentage of defecations occurring at the watering point, resulting in mineral transport from the grazing area to the resting area. Phosphorus is particularly prone to this problem.

Research at the Forage Systems Research Center compared manure distribution by grazing beef cattle at several levels of management intensity and travel distances to water. If cattle are kept within 600 to 800 feet of water and shade is not available, about 80-90% of the manure will be returned to the productive grazing areas. With limited shade available, the amount may be reduced by about 20% as the cattle spend more time concentrated around the limited shade area. Agroforestry systems or natural savannas with abundant shade are probably more efficient for nutrient distribution but there is virtually no research in the area to verify that assumption.

High stock density enhances manure distribution. A 1-2 day rotation frequency with stock density between 25,000 and 40,000 pounds animal liveweight/acre resulted in about 45% of the pasture area receiving dung excretion on an annual basis. Compare this to the 2 to 5% common to continuous grazing and it is easy to see why MiG can significantly reduce the needs for purchased fertilizer. Another important consideration in nutrient cycling management in grazing operations is feeding supplemental feed. Mineral nutrients contained in the feed contribute to pasture fertility as long as the manure is deposited on the pasture.

For grass-based dairies looking for nutrient efficiency, minimizing the time cows are in the parlor and off the paddocks is important. Manure deposited at the parlor is a liability, manure deposited on the paddock is a benefit. In beef operations, feeding hay on the paddocks rather than in a dry lot is a useful way of raising or maintaining pasture fertility. Purchased hay can serve a dual role as feed and fertilizer if fed in the right location. Managing nutrient cycles is just one more part of the "Management" in Management-intensive Grazing. Serious graziers who understand the processes earn substantial savings from reduced fertilizer costs and enhanced sustainability.

The Basics:

- Animals grazing on pasture return over 90% of the nutrients they consume back to the soil through their urine and feces.
- Nutrients flowing through urine have a better distribution pattern in the pasture and nutrients flowing through feces tend to concentrate at animal rest spots.
- Shade, water, and supplemental feeding sites are manure magnets.
- The fewer water sites and the farther between them, the less efficient the nutrient cycle. Conversely, the more watering sites and the closer they are, the more efficient the cycle becomes.
- High stock density enhances manure distribution.

Think About Your Farm or Ranch:

- Take a pasture walk to determine where your nutrients are located. With basic understanding of where the minerals flow, plan how to manage nutrient cycles in your pastures.
- What can you do to change shade, water, and supplemental feeding sites?

Growing Quality Pasture

10 What Factors Make Quality Pasture?

Stage of maturity is the single most important factor determining overall forage quality.

In 1994 I had the privilege of attending a national conference on forage quality at Lincoln, Nebraska along with several hundred professional colleagues. From that conference sprang a thousand-page book dealing with all aspects of forage quality. This is not to downplay the significance of the hundreds of research studies and thousands of scientist years that went into that publication, but we can boil down that thousand pages to a fairly simple statement. A quality forage has low fiber, high protein, and tastes good. The thousand pages are needed to explain why this is not always true.

At some point in its life cycle, almost any plant will meet the low fiber and high protein qualification. However, the tastes good part can be a real problem for some plants. When plants contain chemical compounds, which impart bad tastes or create sickness in animals, we call those anti-quality factors. Endophytic fungus in tall fescue and tannins in serecia lespedeza are two of the best known examples of anti-quality factors. We will talk about anti-quality factors in detail in a future chapter. For now, we will deal with the first parameter, low fiber and its effect on dietary energy level.

Grazing ruminants derive dietary energy from two sources within the plant. The cell contents, or cytoplasm, contains a limited amount of energy in the form of soluble sugars, organic acids, or other rapidly digested compounds. The sugars are the immediate product of photosynthesis which have not yet been incorporated into plant structure or are being used in metabolic processes. If you have read or heard reports of hay cut in the afternoon being higher quality than hay cut in the morning, these cell solubles are the reason why.

Photosynthesis is producing sugars at a rate faster than the plant can utilize them, resulting in more rapidly digestible energy being present in the plant during the peak sunlight hours. Even monogastrics can utilize the readily available cell solubles. The less mature the plant, the greater proportion of total dry weight of the plant made up by cell solubles. This is good.

For ruminants the majority of energy from forage comes from rumen fermentation of fiber. This is where ruminants have their great advantage over monogastrics in utilizing bulk plant material. The main fiber constituents in plants include pectin, hemicellulose, and cellulose. Each of these vary in degree of digestibility and proportion in the plant depending on the stage of maturity with pectin being almost entirely digestible and cellulose being least digestible.

Stage of maturity is the single most important factor determining overall forage quality. As plants become more mature, the digestibility of both hemicellulose and cellulose declines. In immature plants, as much as 90% of the cellulose may be digested while in fully mature grass stems as little as 25% may be digested.

Other non-fiber compounds exist in the plant structure, which can limit digestibility of the fiber components. Lignin is the most important and best known non-fiber component of plant cell walls. Lignin is essentially completely indigestible at any stage of maturity, but lignin concentration increases as plants mature. Compounding the problem is that lignin becomes closely bound to the fiber components with increasing plant maturity, which further limits the digestibility of the fiber by rumen microorganisms. Managing plant maturity to minimize lignin accumulation is one of the fundamental strategies of grazing management.

Another factor affecting lignin accumulation is temperature. Forages at comparable maturity stages will be less digestible in midsummer than in the spring or fall. This is one reason why midsummer steer performance is often disappointing even when pastures look to be immature and very nutritious. This problem is accentuated in legumes compared to cool-season grasses.

Most of what has been said regarding fiber and energy also relates to crude protein content of the plant. As plants mature, crude protein concentration declines as does the availability of the protein. Lignin also binds with proteins and restricts protein utilization by ruminants.

Numerous other factors also help determine forage quality, but plant maturity tends to be the overriding factor in grazing situations. Plant species, fertilization, burning, rainfall, day length, and grazing management are all factors to be considered. Next we will discuss how grazing management can be used to control forage quality.

The Basics:

- A quality forage has low fiber, high protein, and tastes good.
- The sugars in a plant are the immediate product of photosynthesis which have not yet been incorporated into plant structure or are being used in metabolic processes.
- The less mature the plant, the greater proportion of total dry weight of the plant made up by cell solubles. This is good for grazing.
- In immature plants, as much as 90% of the cellulose may be digested while in fully mature grass stems as little as 25% may be digested.

- Managing plant maturity to minimize lignin accumulation is one of the fundamental strategies of grazing management.
- Forages at comparable maturity stages will be less digestible in midsummer than in the spring or fall. This problem is accentuated in legumes compared to cool-season grasses.

Think About Your Farm or Ranch:

- Do your pastures look good, yet result in disappointing animal performance?
- Are you managing plant maturity properly?

11 Creating Quality Forage in Your Pasture

The general rule is:
the higher the nutrtional needs,
the shorter the rest period.

In making high quality forage, remember that high quality forage is low in fiber and high in crude protein, and that plant maturity is the single most important factor in determining forage quality.

The challenge we face in day-to-day grazing management is how do we create that high quality forage on an ongoing basis? We need to think about managing for quality along two different time lines. The first time line is what goes on in an individual grazing period or a series of grazing periods while the second takes place over many years.

One of the first steps in providing quality forage in each grazing period is planning for rest periods of the appropriate length. Once individual plant components reach a certain age, fiber amount increases and fiber digestibility decreases. Regardless of the plant, whether it be bermudagrass or alfalfa, rest periods that allow the plant to reach physiological maturity result in low quality forage. So the 64 dollar question is how long should the rest period be? And the universal answer by any technical advisor is: It depends!

It depends on what plant or mix of plants we are talking about, it depends on the season of the year, and it depends on animal requirements. The appropriate rest period for tall fescue is less than that required by smooth bromegrass regardless of season. Rest periods in rapid growth seasons are always shorter than rest periods during stressful growing conditions. We often talk about fast rotations during fast growth conditions and we say we do it to keep the pasture from getting away from us. The pasture really isn't going anywhere, it just kind of lays there. What we are trying to keep under control is forage quality as much as anything.

The one "it depends" that many people forget about, particularly agronomists, is what are the animal requirements? In the beef business, the classic example we use is stockers vs cow/calf pairs. As a general rule, stockers have higher nutritional requirements than do cows, so rest periods for stocker programs are typically shorter than cow-calf rest periods. Now, we could get into arguments about relative milking ability of cows and whether calves are allowed to creep graze and a host of other contingencies, but the general rule is: the higher the nutritional needs, the shorter the rest period.

During the actual grazing periods, we make decisions that don't alter the forage quality in the pasture but do affect the nutritive value of what the animal consumes. The youngest part of the plant is the highest quality, which is why animals prefer to graze the tops of plants. How much grazing pressure we apply determines how deep into the canopy the animals will graze. Grazing pressure has been replaced by stock density in most grazier's vocabulary. High stock density for a given amount of time will cause the animals to graze deeper into the canopy than would a lower stock density for the same amount of time. Again, as a general rule, higher stock density in any given grazing period will result in lower quality forage being consumed. But, we have a confounding factor we need to consider.

Applying high stock density repeatedly results in higher quality regrowth than does repeated grazing at low stock density. So while the animal grazing at high stock density bites deeper into the canopy, the lower part of the canopy is higher quality than many pastures you see up and down the road. High stock density applied early in the season is one of the important tools used by graziers to create high quality pasture later in the season.

Two practical approaches for achieving high stock density are reduction of paddock size with more frequent moves or increased stocking rate on the pasture during the rapid growth phase and reducing stocking when growth rate slows. The reason we clip pastures or harvest hay or silage when pastures are out of control is to apply excessive stock density to correct the quality problem. That stock just runs on diesel fuel and depreciation.

The second time line is management over the years and this pertains more to what grows in the pasture. Applying high stock density and rest period management are important tools here also, but the goal is to alter plant community, not individual plant composition.

In tall fescue country, many people believe that fescue is an overpowering grass that simply chokes out everything else in the pasture. Fescue is overpowering only if we allow it to be. With high stock density, proper timing of grazing, and appropriate rest period management, what appear to be solid stands of fescue can become very nice, diverse pastures in a few years. The reappearance of native tall grass prairie species in what were thought to be solid stands of fescue is a common site during pasture walks at MiG farms.

Many pastures at the Forage Systems Research Center at the University of Missouri in Linneus, which were 90% tall fescue when I first came there have been reduced to 30% fescue with the other 70% now being a myriad of other grasses and legumes. All as a result of changing basic grazing management.

High quality pastures don't just happen. They are the product of long term planning and timely management. Soil fertility, interseeding other species, fire, mechanical intervention are all tools which can help you along in your quest for the perfect pasture, but all of those inputs are wasted if proper grazing management doesn't come first.

The Basics:

- Remember, high quality forage is low in fiber and high in crude protein.
- Plant maturity is the single most important factor in determining forage quality.
- One of the first steps in providing quality forage in each grazing period is planning for rest periods of the appropriate length.
- The proper length of a rest period is affected by the mix of plants in the pastures, the season of the year, and animal requirements.
- Rest periods which allow the plant to reach physiological maturity result in low quality forage.
- Rest periods in rapid growth seasons are always shorter than rest periods during stressful growing conditions.
- When we talk about fast rotations during fast growth conditions and we say we do it to keep the pasture from getting away from us, what we are trying to keep under control is forage quality.
- As a general rule, higher stock density in any given grazing period will result in lower quality forage being consumed.
- Applying high stock density repeatedly results in higher quality regrowth than does repeated grazing at low stock density.
- High stock density applied early in the season is one of the important tools used by graziers to create high quality pasture later in the season.
- Two practical approaches for achieving high stock density are reduction of paddock size with more frequent moves or increased stocking rate on the pasture during the rapid growth phase and reducing stocking when growth rate slows.

- High quality pastures are the product of long term planning and timely management.

Think About Your Farm or Ranch:

- List the diversity of plants growing in your pastures for each season of the year.
- What is your peak growing season?
- When is your flat spot? What can be done to extend the growing season during this period?
- What class of animals are you using? What are their nutritional needs?
- What are your long term management goals?
- What else can you do to create high quality forage on a continuous basis?

12 Anti-Quality: What Makes Good Forages Go Bad?

Anti-quality factors are in pastures all around us.

A little over 70 years ago in a hill country pasture in eastern Kentucky someone noticed a different kind of grass that seemed to survive summer drought and stay green well into winter. A University of Kentucky agronomist collected some plants and began working with this new grass. It was easy to establish and persisted well on a wide range of soils. It was productive and responded very well to nitrogen. Laboratory tests showed it to have high forage quality. After several years of testing, the new wonder grass was released as Kentucky 31 tall fescue.

Not long after its release, reports started coming back of lameness in cattle grazing the new fescue. More and more cows were coming up open and calves were stunted little hair balls. What could possibly have gone wrong with such an excellent forage?

All sorts of possibilities were investigated, from leaf molds to plant-produced alkaloids. It took about thirty years of hard work before researchers at the University of Georgia identified a fungus growing inside the plant as the causal agent of fescue toxicity. The fungus and plant together were producing highly toxic alkaloids. Chemical compounds that offset much of the good aspects of a particular forage are what are termed anti-quality factors.

Endophyte-infected tall fescue is the classic example of anti-quality that farmers and ranchers east of the Great Plains are most familiar. Estimates of the annual cost of fescue toxicity to the cattle industry range from $400 to $800 million annually.

There are a surprising number of other anti-quality components occurring in a wide range of common, and not so common, forages. Many varieties of perennial ryegrass, which is touted as the highest quality grass in the world, contain toxic alkaloids produced by their own set of endophytic fungi. As with fescue, there are endophyte-free perennial ryegrass cultivars available. Curiously, orchardgrass contains endophytes that seem to produce no toxic compounds.

Older cultivars of reed canarygrass have long been known to contain alkaloids causing diarrhea in sheep and cattle. It had always been assumed they were natural plant products, but now researchers are checking to see if these also might be endophyte products.

Nothing in nature happens without a reason, so why do some plants have endophytes? Self defense is the best explanation for anti-quality factors. Besides producing livestock toxicity, endophytes provide other chemical compounds that protect the plant against insects and diseases. The fescue endophyte has also been shown to enhance the drought hardiness of the plant. In return for protection, the plant provides a home and food for the fungus.

Other common anti-quality factors include tannins, terpenes, cyanogenic glucosides, and oxalates. All of these compounds can be detrimental to grazing animals, but at the same time provide certain plant benefits. There are thousands of plant-produced chemicals that can have toxic effect if certain environmental or animal health situations occur.

Tannin is most often associated with sericea lespedeza, an introduced forage legume now playing havoc with a lot of rangeland in the Plains region. High levels of tannin cause animals to drastically reduce their forage intake, thus protecting the plant from overgrazing. A number of years ago, Auburn University released a low-tannin cultivar of sericea called Lotan in an attempt to improve palatability.

So, are tannins always bad? Actually, annual lespedeza also contains tannin, but at a lower level than sericea. Tannin can protect protein from rumen digestion making it into an effective by-pass protein that can be utilized more effectively from the small intestine. Non-bloating legumes almost all contain tannins at varying levels. There is strong evidence that tannin is the basis of the non-bloating aspect of those legumes. So a little bit of tannin can be a good thing while too much tannin is a bad thing.

Terpenes are commonly found in sagebrush and many other arid land shrubs. Like tannin, terpene sickens the animal and reduces forage intake. In recent years, Dr. Fred Provenza at Utah State University, has found that combining both tannin and terpene in the diet results in less detrimental effect than if animals are exposed to only one of the toxins. Intake of high tannin plants increases if the animals can also consume terpene-containing plants at the same time.

Cyanogenic glucosides are precursors to cyanide, one of the most toxic poisons known to man. Birdsfoot trefoil, a wonderful forage legume, contains the cyanide precursor. Animals rarely consume enough trefoil to ever have a problem. Multiple stress factors on both the plant and animal can combine to create a cyanide occurrence once in a great while.

Oxalates are precursors for oxalic acid, another mild poison, and are commonly found in both grasses and broadleaf plants. It is safe to say, virtually all plants contain some potential anti-quality factor. The issue is dosage and how the anti-quality factor interacts with all other components of the diet. Even our fresh, home-grown garden vegetables contain deadly poisons. Obviously we can't live in constant fear of us or our livestock dying from what we eat so we must move ahead.

The most common way we deal with anti-quality factors in day-to-day management is dilution of the toxin to reduce the dosage. Why do we interseed legumes into endophyte-infected tall fescue? First, to dilute the toxin. Second, to raise the overall nutritional state of the animal so it can better cope with poisons in the diet. As long as range cattle have something else to forage on, they can browse a little sagebrush and be okay.

In the longer view, plant breeders have worked for many years to lower toxicity levels in plants. That is why most of us can stand to eat broccoli. Someone started taking oxalates out of it a long time ago.

Endophytes were removed from both perennial ryegrass and tall fescue so that we could have non-toxic endophyte-free varieties. They were great for the livestock but persistence was lousy in some environments. Remember, nothing in nature happens without a reason. Those endophytes were there to protect the plant from other pests besides just cattle. Now they have put endophytes back into ryegrass and fescue to provide insect and disease protection without causing animal toxicity.

Anti-quality factors are in pastures all around us. We should be aware of their potential and know how to deal with them. As a last resort we can seek alternative forages to replace the most toxic plants.

Anti-quality was best summed up by one of my professional colleagues who said you can feed your children the most nutritious food available but if you sprinkle a little cyanide over it everyday, they just won't do very well.

The Basics:

- Chemical compounds that offset much of the good aspects of a particular forage are what are termed anti-quality factors.
- Self defense is the best explanation for anti-quality factors.
- High levels of tannin cause animals to drastically reduce their forage intake, thus protecting the plant from overgrazing.

- There is strong evidence that tannin is the basis of the non-bloating aspect of those legumes.
- Research has found that combining both tannin and terpene in the diet results in less detrimental effect than if animals are exposed to only one of the toxins.
- Multiple stress factors on both the plant and animal can combine to create a cyanide occurrence once in a great while.

Think About Your Farm or Ranch:

- How can your grass management decisions affect the anti-quality factors in your pasture's forage?

13 Understanding the Value of Plant Diversity

Diversity increases the likelihood that there will be something green and growing more days of the year.

There is a lot of talk these days in both political and scientific circles about the value of biodiversity. We are inundated with horror stories of disappearing species and the gaps their passing leaves in the ecosystem. We are not going to tackle global issues of vanishing species, but we will look at the role of plant species diversity in pastures. Very simply put, the value of diversity is increased likelihood that there will be something green and growing more days of the year.

Let's start by breaking plants into what ecologists term "functional groups." Depending on your level of interest, functional groups can be very broad or fairly narrow. We will start with eight groups: cool-season perennial grasses, cool-season annual grasses, warm-season perennial grasses, warm-season annual grasses, perennial legumes, annual legumes, cool-season forbs, and warm-season forbs. Each of the grass classifications could be further broken down into bunch type vs sod formers and the legumes could also be broken into cool-season vs warm-season and erect vs vining. If you start thinking about plants in those terms, it quickly becomes apparent that each functional group, and perhaps each species, occupies a unique niche in the ecosystem. The native tall grass prairies and their massive herds of large ruminants of the central USA are often touted as a sustainable model for modern grazing management. All eight of the functional groups we described occurred in the native prairie. Between 200 and 300 species of plants typically occurred in an acre of prairie. While all of these species compete for space and nutrients in the community, they also tend to complement one another.

Rather than all growing at one time or trying to germinate new seeds at one time, they have evolved with different photosynthetic pathways and different growth requirements, which gives us as grazing managers that desirable situation of something green more days of every year.

Let's think about some of the complementary relationships we see in solar energy capture. The leaves of most grasses are erect to semi-erect. If you put a light meter at the bottom of a grass canopy of appropriate grazing height, you would find a fairly high percentage of incoming solar radiation reaching the soil surface. This is particularly true in thin grass stands.

This is actually beneficial for cool-season grasses because they cannot utilize full sunlight anyway. It allows more light to reach lower leaves, but it is also a waste of energy if we think in terms of total primary production in the ecosystem. Legume leaves tend to be more horizontal and are more efficient at intercepting more solar energy at a lower leaf area index (LAI) compared to grasses. But when legume leaves are inter-layered through the grass canopy, the system becomes more efficient than either functional group by itself.

Another wonderful compatibility is the seasonal succession of cool-season to warm-season species. Cool-season grasses begin to lose their leaves as summer heat and drought progress, as then more sunlight reaches the lower part of the canopy where the warm-season grasses are just beginning to grow. Warm-season grasses, in contrast to cool-season grasses can utilize full sunlight and are suppressed by shade. In the late summer and fall the opposite process happens as warm-season grasses begin to die back, more light becomes available for cool-season species.

Similar competition occurs below ground as the roots compete for water and nutrients. Some grasses are shallow rooted while others are deep-rooted. Grasses are fibrous rooted while legumes tend to be more taprooted. While almost all forage plants have the most active roots in the surface four to eight inches, some roots may be many feet deeper than others. In fact deep-rooted species such as alfalfa and the native tall grasses have been shown to bring both water and nutrients from deep in the soil profile and make them available to shallower rooted species.

One of the most famous of all plant associations, which points to the value of biodiversity is nitrogen fixation through the symbiotic relationship between host legume plants and the bacteria in their nodules. As the legume roots die back, nitrogen is released to the soil and can be taken up by the grass. In the absence of legumes, N must be provided to the pasture from some external source. Purchased N fertilizer comes at a cost that many producers cannot afford while high levels of manure can cause serious environmental problems.

A less well known feature of plant species diversity is the damper that it places on the spread of harmful insects and diseases. An example comes from work conducted at the University of Missouri by Drs. Gene Munson and Wayne Bailey which looked at the impact of grass in alfalfa hay fields on alfalfa weevil and potato leafhopper infestations. As little as 10% orchardgrass in the sward significantly reduced populations and rate of spread of both insect pests. Numerous other examples can be found throughout the natural world.

Not only does the plant community benefit from plant biodiversity, but so does the animal community. Greater plant diversity broadens the nutritional opportunities of grazing livestock through several avenues. Having more days of something green and growing translates directly to more even distribution of adequate protein, energy, and minerals through the season. Because of the different rooting depths and structure, plant species also tend to have a wide range in mineral composition.

We could go on about wildlife and songbirds and a myriad of other benefits, but the benefits of just the pasture and domestic graziers should be enough to keep everyone positive about biodiversity. Remember green and growing more days makes more profitable grazing.

The Basics:

- Inter-layering legumes throughout the grass canopy creates more efficient pasture than either grasses or legumes alone.
- Warm-season grasses, in contrast to cool-season grasses can utilize full sunlight and are suppressed by shade. In the late summer and fall as warm-season grasses begin to die back, more light becomes available for cool-season species.
- As the legume roots die back, nitrogen is released to the soil and can be taken up by the grass, making this one of the most famous of all plant symbiotic relationships.
- Plant species diversity suppresses the spread of harmful insects and diseases.
- Plant diversity broadens the nutritional opportunities of grazing livestock.
- Different rooting depths and structure create plant species with widely varying mineral composition.

Think About Your Farm or Ranch:

- Take a pasture walk. Note the variety of grasses, legumes and forbs in your fields. Identify your warm-season and cool-season species, perennials and annuals. Chart their growth cycles over a calendar year. Identify your flat spots and lush periods and times when they overlap. How can you add diversity?

14 Working with Legumes in Your Pastures

Some management styles are more compatible with legumes than are others.

I have often referred to grass-legume mixtures in previous sections as an important part of the pasture plan. A common concern for many graziers is how do you get those legumes growing in the pasture in the first place and then keep them there. With the widespread climatic diversity of North America, it is really difficult to address everyone's individual environment and needs as it relates to using legumes.

One of the first principles to accept is some soil types and environments are much more conducive to growing legumes than others. Likewise, some management styles are also more compatible with legumes than are others. And, there are some situations where N-fertilized grass is actually the best pasture alternative, but today we're talking about legumes.

The ideal legume environment is a temperate climate with deep, fertile soils with near neutral soil pH. I guess that is near ideal for almost everything we try to grow and not many of us really seem to have just the right conditions so we have to make do with what we do have.

There are a few basic concepts we do need to be aware of when planning on introducing legumes into our pastures and relying on them for significant contributions.

First, temperate legumes are easier to manage and maintain than are tropical and sub-tropical legumes.

While some legumes can be very competitive on poor soil sites, most tend to perform best on pretty good soil. If you live in a temperate zone, you already have a leg up.

Second, most legumes do not have outstanding seedling vigor and can't stand a lot of competition so it is critical to control existing vegetation during the establishment phase.

Third, grasses are much more efficient at using soil nitrogen than are legumes. If there is an abundance of N in the soil, grasses will thrive and legumes will suffer. With these things in mind, we can start planning on legume use.

In the Midwest and Northeast region, a number of different legumes are available to work with and, fortunately, we can usually come up with something to work in almost all situations.

For lower fertility sites, birdsfoot trefoil and annual lespedeza work well. For wetter soils, white clover and alsike clover are good choices. Alfalfa performs very well on the best sites and red clover works on just about everything in between.

In the Upper and Middle South, a number of winter annual legumes can be used to complement winter annual grass pastures and can be overseeded on warm-season grass pastures.

Crimson and arrowleaf clover and several vetch species can provide several months of grazable forage between fall and spring. red clover and white clover are good choices for permanent pasture. On the Gulf Coast, several tropical legumes such as perennial peanut and joint vetch can be used in pastures. In each case, as we move farther south the challenges of growing legumes increases as insect and disease pressures mount and soil quality declines.

Alfalfa is the primary legume for grazing in improved dryland pastures in the Plains region. A number of native legumes are important components of native range on the Plains, although there are also several toxic legumes.

In the irrigated Intermountain West and the Pacific Northwest, most of the same legumes that are used in the rainy Midwest and East are readily used for grazing. Most locations across the country have legumes that will provide good grazing and make significant N contribution, with proper management. There are several ways to determine which legumes may work in your environment and with your management style.

The best way is by finding the good graziers in your neighborhood and find out what is working for them. On similar soils and with similar management, those species or varieties are likely to work for you also.

University variety trials are helpful but not many universities do more than just alfalfa variety trials anymore. Just finding what is working in university grazing trials is also valuable. Some people have the attitude that what happens at a university farm or research center isn't representative of the real world and doesn't mean anything. After 20-plus years working at a university research center, I can assure you that pastures often face greater challenges in the university system than you can imagine.

The third way to find out what will work for you is the most expensive but most effective route. That is do your own research and try several legumes of interest in your pastures and see how they perform. Most importantly, if you do your own research, keep good records of what you have done.

Fertility is an important issue for legume establishment and persistence. Disappointment in legume performance on many farms can actually be attributed to inadequate fertility management. Low soil phosphorus level is a problem throughout the eastern half of the USA and most legumes thrive only with adequate P. One of the really nice things about P fertilization is that it is a long term investment that can work for you year after year in a well-managed grazing system.

Livestock don't remove much P in the weight gain and milk sold of the farm or ranch, but they move a lot of it around in their manure. That is why water location and stock density management are so critical to keeping P evenly distributed across the landscape in pasture-based production systems.

Before embarking on a serious legume-based pasture program, get the soil tests and deal with fertility issues up front. The results will be much more satisfying.

So, what's the kicker about "with proper management"?

Most of the legumes requires a managed rest period for persistence. For some it is an opportunity to recharge storage carbohydrates while for others it is an opportunity to reseed. There are a few legumes that do alright in continuously grazed pastures, but not many.

White clover and kura clover are able to survive and are reasonably productive if set stocked at appropriate stocking rates.

Even though the newer grazing alfalfa varieties like Alfagraze are touted as being selected under continuous grazing, they perform much better with rotational grazing.

Rest period requirements vary with species and time of year. The best way to determine when and how much rest is needed is by watching the pastures. This is not something you learn overnight or even in a single season.

Over time you learn to watch the trend in species population and find a level of legumes that meets your objectives. The real trick is figuring out how you got the pasture to that condition and trying to repeat the process.

A lot of people have asked me what the most important characteristic of a good grazier is and my reply is always, "Learn to be a good observer." Figuring out why things happen the way they do is half the battle. I guess that's true of life and everything else.

The Basics:

- Some soil types and environments are much more conducive to growing legumes than others.

- Temperate legumes are easier to manage and maintain than are tropical and sub-tropical legumes.
- Most legumes do not have outstanding seedling vigor and can't stand a lot of competition so it is critical to control existing vegetation during the establishment phase.
- Grasses are much more efficient at using soil nitrogen than are legumes.
- Legumes require pretty good soil.
- Phosphorus can be a limiting factor.
- A rest period is critical.

Think About Your Farm or Ranch:

- Do your own research and try several legumes of interest in your pastures and see how they perform. Most importantly, keep good records of what you have done.
- Locate the good graziers in your neighborhood and find out what is working for them.

15 Interseeding Grasses and Legumes

Through diversification a pasture is more likely to have some plant green and growing, producing quality livestock feed on more days of the year.

I've talked about the importance of pasture diversity for creating a more effective solar panel. A basic goal of grazing management is to harvest as much solar energy per unit area as possible to maximize the production of quality livestock feed. Grass leaves are typically vertical or steeply inclined allowing sunlight to penetrate deep into the canopy. Much sunlight can reach the soil, wasting the food potential of that sunlight. By adding legumes with more horizontal leaves or grasses with softer leaves, another layer of solar energy traps cover every acre.

Another reason for diversification is that different species have different growth requirements, some plants grow better at one time of the year than do others. Cool-season and warm-season grasses are the most obvious contrast. Through diversification a pasture is more likely to have some plant green and growing, producing quality livestock feed on more days of the year. This goes back to the basic premise of capturing more solar energy per acre.

While there are times when complete renovation of a pasture and reseeding are appropriate strategies, more often the existing pasture will respond very well to a change in management. Introduction of legumes into grass dominant pastures is probably the most common form of species diversification, however cool or warm-season grasses can be introduced into grass swards. The following deals with the nuts and bolts of making pasture interseeding work.

Controlling Existing Vegetation:

For new grass or legume seedlings to become established in grass sods, the existing vegetation must be controlled. Competition exists both above and below ground and should be managed at both levels. Above ground competition is for physical growth space and sunlight while below ground competition is for water and soil nutrients.

The existing sod should be managed with consideration for both of these zones of competition. Simply keeping the grass "short" may not be adequate if a very vigorous root system still exists below ground. Forward planning to weaken the existing root system is a critical management component to legume establishment. Existing vegetation may be managed through grazing, mowing, tillage, burning, chemical application, or a combination of practices.

Grazing:

If only grazing is to be used for vegetation control, timing is critical. Assuming a spring legume interseeding, heavy grazing pressure should be applied the previous fall and through winter if necessary. The strategy is to weaken the grass root system in order to reduce root vigor and competition for water the following spring.

Grazing a vigorous sod only in the spring may be adequate for reducing above ground competition for sunlight but may be totally inadequate for minimizing below ground competition. Grazing pressure should be maintained in the spring until the legume seedlings have emerged and produced three or four true leaves. Once legumes have reached this stage, controlled grazing should be employed to periodically top graze the sward to keep the canopy fairly open allowing sunlight to reach the legume seedlings.

Grazing management for establishing a warm-season grass into an existing cool-season grass is very similar to management for legume establishment. Newly establishing grasses are not quite as sensitive to grazing pressure as are new legumes due to the growing point for grasses being located at the base of the plant while the legume growing point is near the top of the plant.

Soil moisture conditions should be closely monitored in grazing situations. Cattle should be removed from the pasture if soil moisture conditions are such that the cattle hooves are making more than a one inch imprint in the soil. The damage to new seedling stands caused by grazing when the soil is too wet usually outweighs the stand reduction that may occur if the grass is allowed to become a little more competitive.

Cool-season grasses are more effectively established in an interseeding if the seeding is done in late summer or early fall. Thus, severe grazing through mid-summer is required to prepare the sward for a cool-season grass seeding. Late summer grass interseedings are especially sensitive to drought conditions and fall grazing should be monitored closely if soil moisture is short.

Mowing:

As with grazing, mowing as a tool to control existing sod should begin the fall prior to seeding. Usually mowing in October will cause the grass to use stored carbohydrates and reduce spring growth rate. In the spring, two or three mowings may be required to minimize spring competition. Each subsequent mowing in the spring should leave a little taller residual.

Mowing for hay when the grass is in the boot stage will usually result in good legume regrowth in the summer. Letting the grass hay crop go to full maturity usually results in very poor legume establishment due to the extended period of competition.

Tillage:

In very dense sods such as well established bermudagrass, smooth bromegrass or tall fescue, light tillage may be helpful in ensuring legume establishment. Disking with the gangs set at a minimal angle will provide adequate disturbance in most cases. If soil conditions are good, a field cultivator can be used effectively, but a finishing disk generally gives best results. Disturbing as little as 25% of the sod cover can greatly increase the success rate of interseeding.

For late summer grass interseeding, a light tillage will greatly increase the likelihood of new grass establishment. Late summer thunderstorms can provide very good grass seed coverage in the disturbed sod and result in rapid germination and emergence.

Burning:

An early spring burn just after cool-season grass begins growing can provide good spring vegetation control. Burning too early, prior to green-up, may result in more vigorous grass growth and actually increase competition. Attempting to burn later may result in inadequate burn as too much new green growth prevents the fire from effectively cleaning the field.

As with any use of fire as a tool on pasture or rangeland, make sure you have a plan for controlling the fire and providing rapid fire control response if the fire gets out of control. One other caution, weed infestations can occur following a burn of old established pastures. Be prepared to apply weed control measures if necessary.

Chemical control:

Broad spectrum herbicides such as Gramoxone™ and Roundup™ can be used for chemical renovation of pastures. Existing sods may either be temporarily suppressed by using lower rates of herbicide or may be virtually eliminated by using higher rates. Chemical suppression of the existing sod is particularly useful in dry years. Reduction of the evapotranspiration rate will often determine the success or failure of legume interseedings when moisture is limiting.

A single application of 1 to 1.5 qt of Gramoxone™ or 1 qt of Roundup™ shortly after spring green-up will usually provide adequate vegetation suppression for legume establishment without eliminating the existing grass stand. Spraying and using a no-till drill provides a very effective combination.

Seeding effectively:

In order for seeds to germinate, they must be in good contact with the soil or other growth media, such as dung piles or decaying plant material. Soil-seed contact can be achieved by several methods.

Frost seeding:

Legume seeds are typically fairly dense and will readily work into the soil given several freeze-thaw cycles. The limiting factor regarding what species can be successfully frost seeded is tolerance to cold weather while in the seedling stage. Red and white clover can be frost seeded with very consistent success due to their high seedling vigor and frost tolerance in the seedling stage. Although quite susceptible to freezing in the cotyledon stage, these clover species become very hardy after the emergence of the first trifoliate leaf.

Lespedeza can be frost seeded successfully because the soil temperature required for germination is substantially higher than for cool-season legumes. Lespedeza generally does not germinate until the risk of a late frost has largely passed.

Alfalfa and birdsfoot trefoil are less tolerant of cold temperatures in the seedling stage and are thus more susceptible to stand failure when frost seeded. Seedling vigor is also somewhat lower than for red clover and both are much less shade tolerant than is red clover.

Most successful frost seeding will occur if seed is broadcast about 30 to 45 days before the last frost of the season is expected. The seed must come in contact with the soil for frost seeding to be effective. For this reason, one of the vegetation control methods described above must be implemented.

Spring grazing will allow some of the seed to be trampled into the soil, if soil moisture is not excessive. If chemical suppression of the sod is to be used, it should be done the previous fall. Waiting in the spring until green growth has begun to apply herbicides makes frost seeding much less effective. Pasture burning to retard cool-season grass growth in the spring generally occurs too late as legume seedlings should already be up and growing by the optimum fire date. However, broadcast seeding immediately into the ash bed following a burn has also been reported to be successful. Sod disturbance combined with frost seeding works quite well as the roughened soil surface is more conducive to good seed-soil contact due to freezing-thawing activity.

If pastures are to be harrowed for manure dispersion, broadcast seeding at the same time often gives better results than simply frost seeding, particularly with birdsfoot trefoil and alfalfa. If a harrow is used, seeding can be delayed until later in the spring after the risk of late frost is lessened. Success rate is generally very high if slight sod disturbance is combined with broadcast seeding and harrowing.

No-till drilling:

A no-till drill can be used to establish any of the common legumes in combination with any of the sod suppression methods already described. Seeding date can be delayed until all risk of frost has passed. The drill ensures very good seed-soil contact and emergence is typically much quicker and more uniform than with broadcast seeding methods. Crops with higher seed price, such as alfalfa and birdsfoot trefoil, can be seeded with much greater confidence of success when using a drill. Drilling should occur just prior to normal green up

Late summer grass interseedings are usually very successful with no-till drilling. The greatest cause for stand failure, however, is drilling too deep. When running grass seed in late summer, some seed should be visible on the soil surface behind the drill to ensure that seed placement is not too deep.

Paying attention to the details of vegetation management and seeding are the keys to making interseeding work.

The Basics:

- Above ground competition is for physical growth space and sunlight while below ground competition is for water and soil nutrients. The existing sod should be managed with consideration for both of these zones of competition.
- Cattle should be removed from the pasture if soil moisture conditions are such that the cattle hooves are making more than a one inch imprint in the soil.
- Disturbing as little as 25% of the sod cover can greatly increase the success rate of interseeding.
- Burning too early, prior to green-up, may result in more vigorous grass growth and actually increase competition. Attempting to burn later may result in inadequate burn as too much new green growth prevents the fire from effectively cleaning the field.
- Legume seeds are typically fairly dense and will readily work into the soil given several freeze-thaw cycles. The limiting factor regarding what species can be successfully frost seeded is tolerance to cold weather while in the seedling stage.
- No-till drilling ensures very good seed-soil contact and emergence is typically much quicker and more uniform than with broadcast seeding methods.

Think About Your Farm or Ranch:

- Identify the flatspots in your pasture season and then determine what species to add to fill in the gaps.
- Decide which seeding method will work best for you and begin planning your pasture improvement program.

16 What About Dragging Pastures?

It depends!

One of the questions that I get asked again and again at producer meetings is: "Does it do any good to drag pastures?"

My usual answer is the universal extension answer for all questions, "It depends!" and that is usually followed by my question, "What are you trying to accomplish?"

Someday when you have a lot of time on you hands, think about all of the things you do during a day and ask yourself what you are trying to accomplish with each task. The results may be a little scary when you find that an awful lot of what you do is "just doing".

What are we trying to accomplish with dragging pastures?

The most commonly stated goal is to spread manure piles to accelerate manure decomposition and enhance nutrient cycling. Covering seed or disturbing a thatch layer are other common objectives. Dispersing manure piles may also lead to more uniform grazing. On some farms dragging pastures is a tool to transfer funds to dependent children in a tax deductible manner.

Does dragging to scatter manure piles actually enhance nutrient cycling? For all practical purposes, no research exists to answer this question.

From high school chemistry or building a campfire, we know intuitively that as particle size decreases, rate of reaction increases. Thus, if manure piles are reduced to manure fragments, they will decompose more quickly. The smaller particles also have greater surface area contact with the reactive surface (i.e. the soil). I believe it is a pretty safe assumption that nutrient cycling on the cow-pie scale is accelerated following dragging.

However, looking at nutrient cycling from a whole-pasture perspective, the benefit is much less definite. If over half of the manure is ending up under shade trees and around watering sites, pulling a harrow around the pasture does very little to improve the overall nutrient cycle. In this latter scenario, other issues in nutrient management need to be addressed before pasture harrowing rises to a level of importance to even consider.

The next question might be, is the rate of increase in nutrient cycling economically meaningful?

I really haven't a clue whether it is or not. I do believe some other aspects of dragging pastures may have economic relevance. The chain or flex type harrow is a very useful tool for bringing overseeded legumes or grass into better contact with the soil. The greater likelihood of the overseeded crop establishing is probably worth the cost of dragging.

In a research study at the University of Illinois Dixon Springs Research Station in the late 1970's, legume establishment in pastures with broadcasting and harrowing was shown to be superior to just broadcasting and equal to no-till drilling, for some legumes. Manure dispersal comes as an added benefit.

Spotty grazing is very often the result of livestock avoiding both fresh and old manure piles. Scattering those piles can lead to more uniform regrowth and less selective grazing. To completely avoid manure-induced spot grazing is virtually impossible with any class of livestock that are in a production mode. Dragging pastures after every grazing period to avoid spot grazing is probably not economically feasible.

Timing of dragging can be fairly critical. Autumn dragging to break piles up going into the winter can result in much more even spring growth on pastures that do not receive nitrogen fertilizer. Nutrients contained in the manure are likely to be back into the soil solution for early spring growth if autumn harrowed.

Manure piles that have dried a few months tend to shatter and scatter very nicely this time of the year. It also ensures that legumes seeds contained in dung piles are more likely to come into soil contact in the spring.

If I were going to spend any time dragging pastures, October and November would be when I would do it. Spring harrowing can either accelerate or slow pasture growth rate. Harrowing prior to or at green-up frequently accelerates pasture growth by disturbing the thatch layer and allowing the soil to warm up more quickly.

We have measured 3° to 5° F higher soil temperature on harrowed areas in side by side harrowed and unharrowed strips in early spring. But if harrowing is delayed too long after early green-up, growth rate can actually be slowed due to damage to tender young plant growth.

Delaying dragging too long in the spring can also result in destroying legume seedling growing on dung piles.

Some concern has been expressed that dragging pastures may increase the likelihood of spreading infection of intestinal parasites to grazing animals. While this may be a concern in some very mild, humid environments, it is generally not considered to be a problem in the Midwest and Upper South.

Manure in dragged pastures dries out very quickly during most of the year (May to October) in the Midwest. Exposing more manure surface area to the sterilizing effects of solar radiation kills most parasites. Simply drying the manure out reduces the likelihood of survival for some organisms.

In cooler, cloudy climates, parasite persistence is much more of a problem. Parasite reinfection due to dragging is also more likely to occur with horses than with cattle due to the very severe overgrazing habits of set stocked horses. We have kicked the piles around and thought about some of the benefits of dragging pastures.

It is very difficult to say whether or not the benefits of dragging are worth the cost of doing it. In terms of economic importance, I would rate the benefits in the following order: reduce spot grazing (enhanced utilization rate is the economic benefit); improve seed/soil contact (the economic benefit is improved pasture productivity; and accelerate manure decomposition (the economic benefit is reduced fertilizer input).

The Basics:

- The most commonly stated goal is to spread manure piles to accelerate manure decomposition and enhance nutrient cycling, but this may actually be the least economically important.
- Dispersing seed from manure piles and evening out regrowth are more economically important results of pasture dragging.
- Legume establishment in pastures with broadcasting and harrowing has been shown to be superior to just broadcasting and equal to no-till drilling for some legumes. Manure dispersal comes as an added benefit.
- Dragging pastures after every grazing period to avoid spot grazing is probably not economically feasible.

Think About Your Farm or Ranch:

- Before dragging pastures, determine the goals dragging will accomplish for you.

Matching Forage With Animals

17 Why Match Forage Supply to Animal Demand?

Both supply and demand curves have highly predictable peaks and valleys.

For thousands of years grass and grazing animals peacefully coexisted without any external management being applied, so why do we need grazing management in the first place? There is a substantial difference between grass and animals peacefully coexisting and a farmer or rancher making a living. That difference is called management.

One of the basic premises of peaceful coexistence is that animal demand not exceed forage supply. In the natural world, starvation and death is the result of such an imbalance reducing stocking rate back to a level that is within carrying capacity of the resource base. In modern agriculture, imbalance of animal demand and forage supply rarely results in starvation. We have several more merciful ways of dealing with the problem by either reducing demand or increasing supply.

The most obvious way to reduce demand is by getting rid of some animals. One of the common circumstances that knocks supply and demand out of sync is drought, but very few ranchers choose to reduce demand by selling animals until prices are depressed due to over-liquidation by all the neighbors. In a cow-calf operation, early weaning calves reduces overall forage demand. But if calves are sold at a lighter weight, overall return may be reduced. A less often considered way of reducing animal demand is accepting a lower level of animal production. One common theme for all of these avenues of reducing animal demand, is that they all come at the cost of reduced income. Not many ranchers are ready to accept that.

If we look at means of increasing forage supply, several options are immediately apparent. Fertilization is a good example. Put nitrogen on and more grass grows. I once heard the definition of an agronomist as someone who never ceases to be amazed that nitrogen makes grass grow. Of course fertilizer comes at a price also, especially these days.

Pastures can be renovated and more productive species sown, which also comes at a price. More pasture can be rented or purchased and forage supply increased, but guess what? It comes at a price. The guy who said, "Hay is for horses, straw's cheaper, grass is free," had no idea what he was talking about. Grass is never free. While forage supply can be increased cost effectively, many times any additional income due to increased supply isoffset by highercosts and mediocre grazing management.

This discussion might make you think that any reduction in animal demand is going to be coupled with reduced income and any increase in forage supply is going to cost money. While this is often true, it isn't necessarily so. Why not? Because animal demand is not constant 365 day of the year and forage supply is also not constant.

Using the very simple model of a cow-calf operation with no other enterprise we find a very predictable annual cycle of nutrient demand. Highest demand occurs 30 to 90 days after calving while lowest demand occurs 90 to 120 days before calving, assuming constant body condition score.

On the grass side, there is usually a dormant season either due to cold or dry conditions. When growing conditions are good, there is a flurry of growth followed by a return to dormancy. There is usually a fairly predictable growth curve based on the plant species present.

In some parts of the country, the period of rapid growth may be as short as a couple of months while in other areas there may be fairly constant growth for several months. The bottom line is that both supply and demand curves have highly predictable peaks and valleys. It is a fairly simple process to time calving so the peaks of each cycle overlap. In this scenario, no income loss need occur and no additional pasture costs were incurred. All it took was a change in attitude and management.

This simple example used one enterprise and a common forage supply. It is actually easier to bring forage supply and animal demand into balance with more than one animal enterprise and a more diversified forage system. Next we will look at more complex forage and animal systems and more long term farm or ranch planning to help maintain long term balance between animal demand and forage supply.

The Basics:

- The most obvious way to reduce demand is by getting rid of some animals.
- Another way of reducing animal demand is accepting a lower level of animal production.
- Adding nitrogen fertilizer can increase short-term forage supply, but it comes at a cost.
- To increase pasture diversity sow more species or change grazing management.
- Animal demand is not constant 365 day of the year and forage supply is also not constant.
- The simplest solution is to time calving so that the peaks of each cycle overlap.

Think About Your Farm or Ranch:

- If you add fertilizer or seed to increase forage diversity and growth, will the additional income justify the necessary costs?

- Are you willing to change your attitude and management so that your calving and forage peaks occur simultaneously?

18 How to Match Forage Supply and Animal Demand

Flexibility in stocking options is essential to balancing forage supply and animal demand.

In the last chapter we introduced the idea of matching periods of peak nutritional demand by livestock with peaks in forage supply. Using the example of a cow-calf system, energy demand reaches its peak when cows are near peak lactation, which is typically 30 to 90 days after calving and is lowest just after weaning. Depending on the milking potential of the cow, change in energy demand ranges from 30% to nearly 100% (*See diagram*).

A very high milk producing cow at peak lactation is equivalent to two of her at maintenance in terms of effective stocking rate. Put a little more simply, stocking rate can double without adding any more animals.

Spring and early summer forage production may be 100% to 200% greater than mid-summer pasture production. Timing calving to make these two peaks coincide is one of the most fundamental principles when planning a sustainable pasture-based livestock system.

The same principle is the foundation of seasonal grass dairy production. Producing the most milk from the cheapest feed generates the greatest profit. Spring grass is the cheapest feed around because you can do almost nothing and most years

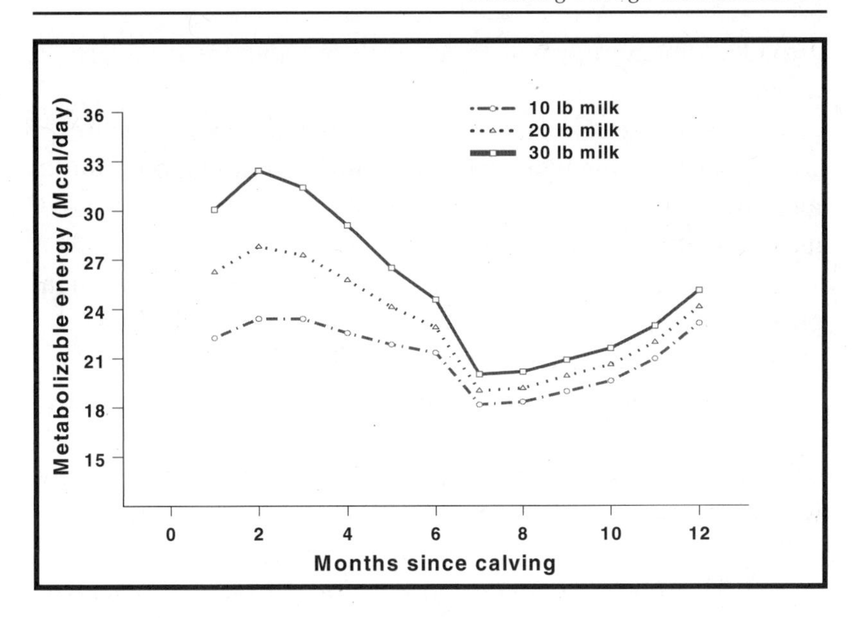

it will be there. Providing pasture any other time during the year usually comes at a higher cost.

So what do we do when animal demand is still high and forage supply is declining? Increasing forage supply during the periods of deficiency is one strategy.

There are several ways of increasing forage supply. Changing grazing management and providing planned rest periods is one means of changing the makeup of the plant community to a more productive, diverse community.

Developing a pasture program that provides the highest annual yield is not necessarily the most desirable management goal. Seasonal yield distribution is equally important, particularly in dam-offspring and dairy systems where animals must be fed on a year around basis.

An ongoing research project at the Forage Systems Research Center has found tall fescue fertilized with 120 pounds of N annually produces about the same amount of forage as tall fescue interseeded with red clover without N fertilizer. But the really key factor is that the fertilized pasture produces most of its forage before July while the grass-legume provided more uniform distribution throughout the summer months.

Warm-season grasses can also increase in the pasture with proper grazing management resulting in even more summer production. In warmer climates, the same statement can be made regarding cool-season species.

While the plant community can be changed just through grazing management, the process can be accelerated with interseeding and strategic fertilization, but always at a cost.

Regardless of what we do with forage supply, it is still very difficult to bring animal demand and forage supply into perfect balance with only one class of livestock. No matter how well we manage our diverse, multi-species plant community, we still seem to have excess pasture in the spring.

The most common solution to this problem of run away pastures is harvesting the excess as hay or silage. Adding another animal enterprise to the system can make managing forage supply much easier by creating a variable stocking rate system.

Here are some of the livestock options we have used on our own farm to try to achieve balance. Our goal is to utilize all of our spring forage by grazing and minimize any hay making, reducing stocking rate in late summer, and maintaining our base cow herd over winter on stockpiled pasture.

The first mixed livestock system we used was a spring cow-calf and spring ewe-lamb operation. With cows calving beginning April 1 and ewes lambing April 15, we were able to put a lot of animal demand on spring forage. Ewes come to peak lactation in about three weeks so they can double their forage intake in a very short time period. Their demand also fades fairly quickly but demand by growing lambs increases until hot weather depresses their appetite and slows their growth rate.

While this mixture allowed us to use a lot of spring grass, it has a couple of basic flaws.

First, all of those cows and ewes are still there in the

wintertime. Second, it takes a lot of sheep to eat much grass. During this phase we also overwintered our own calves and ran them as spring stockers and sold them in late July.

If I had to identify the biggest shortcoming of this program it would be too much winter forage demand.

From this first mixture we changed to spring cow-calf with spring contract stockers kept until early August, liquidating the sheep operation in the process. This mixture worked very well for a few years. In my one ideal year, we made no hay, maintained the cow herd on pasture until February 25, and fed carryover and bought hay for 42 days. I found that if I ran two to three times the number of stockers as I had cows, the balance came out about right.

Yearly variations in pasture conditions obviously dictate the appropriate ratio. Stockers were on the farm from April 1 to August 1. All the pastures used during spring and summer by the stockers were then stockpiled for winter cow feed.

The challenges we encountered were finding a cattle owner who didn't want to leave the steers there until winter, getting any cattle at all on the up side of the cattle cycle when feed prices were low, and maintaining steer gain through July.

Even though we have repeatedly shown that 85% of steer gain in our environment occurs before July 15, most cattle owners want to leave stockers on pasture until October or November because of feedlot finish timing. Not many seem to want to do the 120 days backgrounding program that fits our needs.

More recently we have run dry fall-calving cows on contract as our extra animals. This has worked out very well, as the owner wants the cows back before calving, so he or she has as much incentive to get them off our place by August 15 as we do.

Whereas stockers must be managed as a performance group and in a leader:follower system would lead our cows, the dry fall cows can be used as a true garbage-eater group and run as followers behind our cows. The dry cows are run on a per day charge rather than rate of gain, so we know what our income will be as soon as we know the number we will have. These dry cows run on about a 1 to 1.2 ratio to our cows.

These are just a few examples of livestock options that attempt to bring animal demand and forage supply into balance. Many other combinations can work and may be more appropriate in different environments, both climatic and economic environments. The key point is that flexibility in stocking options is essential to balancing forage supply and animal demand.

The Basics:

- A very high milk producing cow at peak lactation is equivalent to two of her at maintenance in terms of effective stocking rate. Stocking rate can double without adding any more animals.
- Spring and early summer forage production may be 100% to 200% greater than mid-summer pasture production. Timing calving to make these two peaks coincide is one of the most fundamental principles when planning a sustainable pasture-based livestock system.
- Changing grazing management and providing planned rest periods is one means of changing the makeup of the plant community to a more productive, diverse community.
- Fertilized pasture produces most of its forage before July while grass-legume pastures provide more uniform distribution throughout the summer months.
- Warm-season grasses can also increase in the pasture with proper grazing management resulting in even more summer production. In warmer climates, the same statement can be made regarding cool-season species.
- While the plant community can be changed just through grazing management, the process can be accelerated with interseeding and strategic fertilization.

- Adding another animal enterprise to the system can make managing forage supply much easier by creating a variable stocking rate system.
- Yearly variations in pasture conditions will dictate the appropriate ratio of forage to animals.

Think About Your Farm or Ranch:

- What is your goal regarding forage supply and animal enterprises?
- Can you think of three combinations that will work in your environment?

19 Stocking Rate and Carrying Capacity

We must have flexibility in our stocking rate decisions.

One of the most important decisions graziers make is selecting proper stocking rates for their operation. The reason it is so important is because stocking rate affects just about everything in the pasture ecosystem. Understanding the difference between stocking rate and carrying capacity and how they change seasonally and annually is essential to being a successful grazier.

Stocking rate is a measure of animals assigned to a grazing unit for an extended period of time. It may be expressed as the number of animals, animal units, or animal liveweight assigned to a grazing unit relative to land area. Part of the confusion surrounding stocking rate is that any of these three units can be associated with it. Adding to the confusion is that some graziers talk in terms of cows per acre while others use acres per cow, depending on environment.

If you want to say cows per acre, then I need to ask if they are 900 or 1400 lb cows. The animal impact on soil and plant resources is obviously much greater for the larger cow. This is actually how a lot of ranchers got themselves into serious trouble in the last few decades by thinking only in terms of cow units. A ranch that readily supported a cow to three acres in your father's time won't support a cow to three acres for you when his cows weighed 900 pounds and yours weigh 1350 pounds. That weight change is a 50% increase in effective stocking rate even though the ranch has the exact same number of animals on it. If all other factors remained the same, it would take 4.5 acres to support each cow today.

To avoid the issue of animal size, the standard animal unit system was developed. Most Western ranchers are familiar with animal unit months (AUMs) but many of their Eastern counterparts remain clueless about animal units. An AUM is actually an increment of forage use. An AUM is the forage requirement for maintenance of one animal unit for 30 days. But confusion occurs here also as there is some disagreement about the definition of the base animal unit. The most common definition is a 1000 pound dry, pregnant cow at maintenance in a body condition score of 5 on the 1 to 9 scale.

There is another school of thought which uses a 1000 pound lactating cow as the base unit. Fractional units are then used to describe other animal species and classes so that comparisons can be made across species and classes.

The first definition is used by most scientists and government agencies while the latter definition is used by many producers because it is an easy concept to visualize.

And finally we have the use of animal liveweight per acre to describe stocking rate. This approach also seeks to make comparison across species and classes simpler. Unfortunately, 500 pounds of lactating ewes raising triplet lambs does not consume the same amount of forage as 500 pounds of a dry, pregnant cow. This is an easy approach for the mathematically-inclined grazier who likes to make calculations using available forage, utilization rates, and intake estimates. You can make a calculation plugging in values for each of the above terms and come up with an estimated stocking rate as pounds-liveweight/acre. The problem with making calculations is that the grazing system is dynamic and everything is always changing. Because we tend to keep the bad habits we acquire when we're young, I usually use pounds of liveweight/acre when discussing stocking rate. Bear in mind that stocking rate is just the number or weight of critters you decide to put out in the pasture. It may or may not be the proper stocking rate for your resource base.

Another term needs to enter the discussion now: carrying capacity. Carrying capacity is the stocking rate at which animal performance goals can be achieved while maintaining the integrity of the resource base. An important question is whether that target performance level is a production or economic target.

Is carrying capacity of a particular farm or ranch constant?

While Juliet mourned the inconstant moon, we must mourn the inconstancy of carrying capacity. To make matters worse, we must consider both biological carrying capacity and economic carrying capacity, each of which can change year to year. Biological carrying capacity depends a great deal on forage productivity. Year to year variation in rainfall, late or early frosts, cyclicity of certain plant communities all affect the annual carrying capacity. The best we can hope for is to plan to be in the ballpark and make stocking adjustments as necessary.

Economic carrying capacity depends on abiotic factors including where we are in the cattle cycle, the price of milk, and the cost of off-farm inputs and so on. In a particular year, a 30% increase in nitrogen fertilizer price may have a significant effect on the economic carrying capacity of farms with heavy N dependency.

To make matters more challenging, the annual pattern of forage production is also variable. While a cool-season grass-legume pasture might carry 1200 lb-liveweight/acre in May and June, it might carry only 400 in July and August.

What all of this tells us is that we must have flexibility in our stocking rate decisions. Plan on having a variable stocking rate to best match forage resources with animal demand. The greater the seasonal extremities in forage supply, the greater stocking flexibility needs to be.

When visiting with graziers in Canada, I often suggest that they get out of the cow business and graze only stockers. This is a reflection of an environment that can have five months of abundant forage and seven months of snow cover. A 0% to 100% variance in stocking rate might be very appropriate.

Forage demand or effective stocking rate can be varied seasonally without changing animal numbers on the ranch. The additional nutrient demand of lactation provides significant seasonal increase in effective stocking rate. The forage intake at peak lactation can be 30 to 100% higher for a beef cow compared to maintenance requirement. Forage intake for ewes nursing triplets may be 150% above her maintenance intake.

Timing peak forage demand to coincide with peak forage supply is a means of manipulating effective stocking rate without bringing additional stock onto the farm or ranch.

Calving date largely controls seasonal forage demand in a beef cow-calf operation.

Contract stockers, replacement heifers, or boarding cows are all ways of flexing stocking rate through the season as needed. Any system which involves purchase and sale of animals to vary stocking rate seasonally can be a high risk financially. North-South partnerships can be a very effective means for graziers in different parts of the country to increase stocking flexibility while minimizing financial risk.

Managing stocking rate effectively is fundamental to being a successful grass farmer. Next we will further explore the major impact stocking rate decisions have on pasture and animal production.

The Basics:

- Stocking rate affects just about everything in the pasture ecosystem.

- Stocking rate is a measure of animals assigned to a grazing unit for an extended period of time.
- An AUM is the forage requirement for maintenance of one animal unit for 30 days. The most common definition is a 1000 pound dry, pregnant cow at maintenance in a body condition score of 5 on the 1 to 9 scale.
- Stocking rate is just the number or weight of critters you decide to put out in the pasture. It may or may not be the proper stocking rate for your resource base.
- Carrying capacity is the stocking rate at which animal performance goals can be achieved while maintaining the integrity of the resource base.
- Plan on having a variable stocking rate to best match forage resources with animal demand. The greater the seasonal extremities in forage supply, the greater stocking flexibility needs to be.
- Forage demand or effective stocking rate can be varied seasonally without changing animal numbers on the ranch.
- The forage intake for a beef cow at peak lactation can be 30 to 100% higher compared to her maintenance requirement. Forage intake for ewes nursing triplets may be 150% above her maintenance intake.
- Timing peak forage demand to coincide with peak forage supply is a means of manipulating effective stocking rate without bringing additional stock onto the farm or ranch.

Think About Your Farm or Ranch:

- How and when will your stocking rate and carrying capacity change through the seasons?
- How will you measure animal demand?
- Factor in variables such as available forage, utilization rates, and intake estimates to estimate stocking rate as pounds-liveweight/acre.
- Is your performance level a production or economic target?
- Should your carrying capacity be constant?

20 Stocking Rate Affects Nearly Everything

Use of variable stocking rates across the grazing season allows much more efficient utilization of forage resources.

In the last chapter, stocking rate was introduced and defined as a measure of animals assigned to a grazing unit for an extended period of time, expressed as the number of animals, animal units, or animal liveweight assigned to a grazing unit relative to land area. Here we'll look at why stocking rate decisions are so important and consider the many factors stocking rate affects.

The short-term economic ramifications of stocking rate decisions occur with both the buying and selling of livestock. The more animals purchased, the larger the indebtedness and interest payment. If you are using your own capital, greater risk of possible financial loss must be accepted.

A very basic question is: what stocking level can you afford? On the sales side, higher stocking rate means more product to sell. Or does it? It might mean more animals to sell, but there may fewer pounds to sell. Increasing stocking rates usually result in decreasing individual animal performance.

Increased stocking rate means more animal demand on the forage base. Without concurrent increase in forage production or quality, that means reduced intake. The importance of maintaining intake will be discussed.

Always remember, the more you eat, the fatter you get! As intake declines, individual performance declines. The figure on the next page shows the ADG and gain per acre for steers in a 5-year FSRC grazing study. The slopes of the lines can be changed through management but the basic relationship always holds true. While a certain amount of individual performance loss can be traded for increased output per acre, eventually a level of performance is reached which is the minimum requirement to pay the per head costs of each animal. If performance drops below this level, you are losing money.

Overstocking for one year and losing money is a short term economic consequence that may be bearable.

Overstocking year after year has long term economic and environmental consequences that are not sustainable.

In the previous chapter, I used the example of the ranch which used to support one 900-pound cow on three acres comfortably but found it could not support a 1350-cow on the same land base. That was an example of thinking of stocking rate only in a cow/acre mentality. What happens as animal liveweight per acre increases is overuse of the forage resource. As stocking rate slowly creeps up and animal liveweight per acre increases, the forage residual remaining after each grazing and after each season declines. The pasture solar panel is slowly being worn out. Forage productivity spirals downward.

As above ground productivity declines, root growth is steadily cropped back. The pruned roots no longer explore the deeper reaches of the soil profile for water and nutrients, and leaf production is further restricted. By this time soil organic matter is being depleted as well. The life and resiliency of the soil is being stomped out of it as soil structure collapses with loss of organic matter and vigorous roots. But the downward spiral doesn't even end here.

Soil is not held as tightly to the landscape and erosion ensues. The continuing loss of topsoil further erodes the production potential of the land. The water cycle is severely disrupted as infiltration and percolation become more restricted.

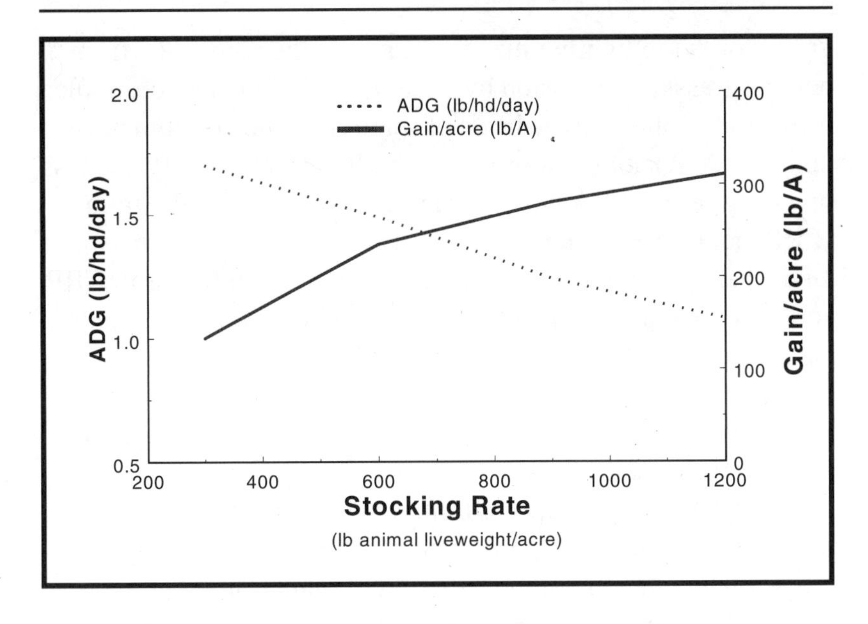

The increased runoff carries fecal coliforms, nitrates, and phosphates and other pollutants into the surface water system and suddenly the state and federal governments are on our back. The realization now comes to the rancher that a cow to three acres doesn't work anymore. Either herd size must be reduced and future income potential lost or a massive infusion of off-farm resources occurs and costs significantly increase, or the neighbor's farm must be purchased and costs increase. These are the long term economic consequences of overstocking. All of this occurred because the manager failed to consider the impact of changing stocking rate on the total production system. What was the manager likely doing at the time he/she should have been thinking this out? Probably reading the latest bull catalog to find another bull with a +50 weaning weight EPD.

While this gloomy discussion may have scared some of you into swearing that you will reduce your stocking rate and avoid this path to destruction, we must also be aware of the dangers of understocking. Grasslands evolved with grazing ruminants over many thousands of years. If a grassland is not grazed or is too lightly grazed, it starts on the path away from being a grassland. Invasion by woody species and less desirable forbs often comes on the heels of removing grazers from grasslands. Low stocking rates usually result in lower quality and less diverse grasslands. A look at Conservation Reserve Program (CRP) acres around the country illustrates the kind of grasslands that exist in the absence of grazing. Many millions of CRP acres are low in plant density and infested with noxious weeds because someone in a place of power didn't understand basic ecological processes..

We should avoid understocking with the same vigilance as we avoid overstocking.

Economically and environmentally sustainable livestock production is a knowledge-based business. There is no substitute for understanding the interaction of plants, animals, and soils. Use of variable stocking rates across the grazing season allows much more efficient utilization of forage resources, but they also allow us to protect the forage and land resource from overuse in times of forage deficiency.

Most of the discussion above focused on the negative effects of increasing stocking rate above the carrying capacity of the forage resource. There are times when overstocking can be beneficial to the overall system. The key is to know when and how much overstocking to use.

Many times pastures which have been under extensive management (or non-management) contain many less-than-desirable species and may also contain much decadent plant material. Short term overstocking of these pastures can actually accelerate their rate of recovery and return them to higher levels of productivity and quality. The increased forage demand associated with overstocking may encourage animals to use the undesirable forages and tramp down the decadent material. The manager just needs to be aware of when to apply the pressure and when to turn it off.

The idea of turning animal pressure on and off leads to the concept of stock density which we will explore next.

The Basics:

- Increasing stocking rates usually result in decreasing animal performance.
- Increased stocking rate means more animal demand on the forage base. Without concurrent increase in forage production or quality, that means reduced intake.
- As stocking rate slowly creeps up and animal liveweight per acre increases, the forage residual remaining after each grazing and after each season declines and the pasture plants are slowly being worn out.
- Long term overstocking depletes soil organic matter, contributing to erosion and disrupting the infiltration and percolation of water.
- Low stocking rates usually result in lower quality and less diverse grasslands, allowing the growth of woody species and undesirable forbs.
- Vigilantly avoid understocking as well as overstocking.
- Variable stock rates also allows us to protect the forage and land resource from overuse in times of forage deficiency.
- Sometimes short-term overstocking can be beneficial.
- Variable stocking rates throughout the grazing season result in healthy pastures.

Think About Your Farm or Ranch:

- What stocking level can you afford?
- Could short-term overstocking benefit your pastures?
- How will you stock your pastures during an ideal pasture growing season?
- How would your stocking decisions change if you faced a drought, extreme winter, or periods of excessive rain?
- What indications in your pastures will determine altering your stocking rate?

21 Tapping into the Power of Stock Density

The key to making stock density an effective tool is knowing when to turn the pressure on and when to turn it off.

We usually think of stocking rate in seasonal terms with long term effects on plants and soils. In contrast, we think about stock density in immediate terms and its effects are readily visible on a daily basis. The concentration of animals on a small area when pastures are subdivided for controlled grazing results in much higher stock density than traditional grazing systems ever provide.

Management of stock density is one of the most important tools graziers have at their disposal. As with stocking rate, plant, soil, and animal condition and performance are all affected by stock density.

Stocking rate is the number of animals or animal liveweight assigned per unit area for an extended period of time. For instance, if we have 100 acres and 50 cows, the stocking rate is one cow/two acres. If the cow weighs 1200 pounds, the stocking rate could also be expressed as 600 pounds liveweight/acre.

Stock density is the number of animals or animal liveweight assigned per acre on a daily basis. In continuous grazing systems, stocking rate and stock density are equivalent because the stock have the opportunity to graze every acre every day. The stock density for continuous grazing in our example would be 600 pounds liveweight/acre.

If the pasture were subdivided to 20 paddocks, we would have all 50 cows on five acres. The stock density is now ten cows or 12,000 pounds liveweight/acre. The stocking rate has not changed but we are now applying 20 times more animal pressure per acre on any given day.

If we were to subdivide to 40 paddocks to facilitate a daily rotation, all 50 cows would be on 2.5 acres. Stock density is now 20 cows or 24,000 pounds liveweight/acre or a 40-fold increase in animal impact. It does not take much imagination to appreciate the effect of increasing grazing pressure by 4000%!

The key to making stock density an effective tool is knowing when to turn the pressure on and when to turn it off. It is animal concentration that results in uniform grazing, consumption of otherwise unacceptable plants, uniform manure distribution and the pulsing of nutrients into the system which make MiG an effective pasture management tool. As a general rule, increasing stock density increases the degree of animal impact. And, as with most things in life, too much stock density can be detrimental.

The keys to using stock density as a management tool are understanding the factors which determine optimum stock density for your particular set of resources and production goals and always thinking of stock density in the context of time. Optimum stock density is determined by four factors, which vary through the grazing season and all of which interact with one another. I am often asked what is the best stock density and I always give my usual answer of: “It depends.”

What it depends on are the four factors shown below in the stock density equation. These are the “Big Four” that apply to pasture or range anywhere in the world. There are some other special considerations for optimal stock density for certain soil types and environments with soils particularly prone to compaction and range sites with certain poisonous plants which are only consumed above a threshold level of grazing pressure.

The formula for figuring stock density is:

$$\text{Stock density} = \frac{(\text{Available forage X Temporal utilization rate})}{(\text{Daily forage intake X \# of days in grazing period})}$$

Available forage is the quantity of forage dry matter allotted to the animals for a grazing period expressed as pounds of forage/acre. Accurately measuring forage availability is time consuming and expensive so we tend to rely on our best estimates of yield. The simplest method is to look at a pasture and make an educated guess as to what the forage availability is likely to be. With practice, a good grazier can consistently estimate within 20 percent± the actual yield. A second method relates height and condition of the pasture to dry matter yield. Tools such as weighted discs, rising plates, and capacitance meters can also be used.

The grazing period or temporal utilization rate is the percentage of the available forage that will actually be consumed by the grazing animals. In the strong continental climates, we usually shoot for 50% utilization rate in most grazing periods. I am a strong believer in the old range adage, "Take half, leave half." It is surprising to many people that seasonal utilization rates in excess of 80% can still be achieved with never removing more than half the standing forage in a single grazing.

Daily forage intake is the amount of forage consumed daily by an individual animal expressed as a percent of bodyweight (e.g. 3%). For the mathematicians who want to make sure all the units cancel in the equation, the units are actually lb-forage/lb-liveweight. To achieve a target level of performance, a very specific daily intake of energy, protein, and minerals are required.

High intake rates can only be maintained at low to

moderate levels of utilization. Exceeding 50% utilization will generally result in decreasing animal performance due to reduced intake.

For example, an average daily gain of 2.25 lb/hd/day is a high performance objective, so utilization can not be excessive or else intake will be limited. To maintain an intake rate of 3% of bodyweight, a 50 percent utilization rate appears to be appropriate to use in the calculation.

Assuming forage availability of 3250 lb/acre and a 3 day grazing period, we can make the following calculation:

$$\frac{3250 \text{ lb forage/ac X .5 utilization rate}}{.03 \text{ lb forage/lb liveweight/day X 3 days}} = 18{,}055 \text{ lb liveweight/ac}$$

If steers weigh 600 lb/head and each acre will support 18,055 lb liveweight, the pasture can be stocked at the rate of about 30 steers/acre/3 day period (18,055 lb liveweight/acre÷600 lb liveweight/steer). A herd of 100 steers require about 3.5 acres/paddock (100 steers÷30 steers/acre).

It is very important that values used for the parameters in the equation are realistic in how they relate to one another. All of the parameters are interrelated and inserting an inappropriate value for any one parameter will result in erroneous conclusions. For example if available forage is below 1500 lb/acre, an intake of 3% would be impossible to achieve. For this reason, the equation cannot be used as most mathematical formulas where if all but one value is known the remaining value can be calculated. A calculation can be made, but the result may be biologically meaningless and, quite possibly, disastrous.

Because our goal is to avoid pasture disasters, a good dose of common sense and understanding plant-animal relationships is always in order. Because stock density is such an important tool for graziers, the concept will keep cropping up.

The Basics:

- Think of stocking rate in seasonal terms, but think of stock density in immediate terms.
- Management of stock density is one of the grazier's most important tools.
- The concentration of animals on a small area when pastures are subdivided for controlled grazing results in much higher stock density than traditional grazing systems ever provide.
- Stocking rate is the number of animals or animal liveweight assigned per unit area for an extended period of time.
- Stock density is the number of animals or animal liveweight assigned per acre on a daily basis.
- Animal concentration or herd effect results in uniform grazing, consumption of otherwise unacceptable plants, uniform manure distribution and the pulsing of nutrients into the system, which make MiG an effective pasture management tool.
- Too much stock density can be detrimental.
- The keys to using stock density as a management tool are understanding the factors which determine optimum stock density for your particular set of resources and production goals. Always think of stock density in the context of time.
- Available forage is the quantity of forage dry matter allotted to the animals for a grazing period expressed as pounds of forage/acre.
- The grazing period or temporal utilization rate is the percentage of the available forage which will actually be consumed by the grazing animals.
- Daily forage intake is the amount of forage consumed daily by an individual animal expressed as a percent of bodyweight.

Think About Your Farm or Ranch:

- On a pasture walk, estimate the amount of forage available. How many paddocks will you need to meet your goals?
- What factors affect the optimum stock density for your resources and production goals?

Managing Pastures and Animals

22 Forage Supply: The Grazier's Checking Account

A big advantage of MiG is the opportunity for the pasture manager to easily assess and keep a record of pasture condition and supply.

Every once in awhile some of us receive an envelope from the bank and inside is a single piece of paper with a big "Overdraft" printed across the top. We seem to think our checking account always has plenty of funds available and we don't need to ever balance the check ledger. Sooner or later we find out we should have been paying a little more attention to our account balance.

Monitoring forage supply is a lot like watching your check book. Most of the time the balance is okay, but if you're not watching it, that overdraft statement can hit you by surprise. The pasture overdraft statement comes in the form of a forced livestock sale or buying in extra feed. In this section we'll look at ways of monitoring forage supply.

A big advantage provided by MiG is the opportunity for the pasture manager to easily assess and keep a record of pasture condition and supply. Continuously grazed pastures are difficult to look at and estimate how many grazing days are available for the herd. When we cut hay, it is easy to count bales to come up with an accurate estimate of how many days of feed we have on hand. Pastures with multiple subdivisions, whether permanent or temporary, are more like the hay field than the uncontrolled pasture when it comes to counting days of feed available. On any given day we can look at each paddock and estimate how many grazing days are available. If we have been monitoring paddocks on an ongoing basis, we also have a very good idea of how fast they are growing and can make fairly accurate projections of future forage supply.

How to actually make those estimates is what bothers a lot of graziers. Part of the problem is that they are unsure what they actually want to know or how they are going to use the record. There are two basic ways of monitoring forage supply. One is animal-based and the other is forage-based.

The first method is actually the simplest, but takes the most experience to do effectively.

In the animal-based approach, everything is expressed on an animal unit or class of livestock basis with no actual estimate of forage dry matter yield ever considered.

I usually think in terms of cow-days/acre because my primary livestock operation is cow-calf. For someone who only backgrounds steers, the term is likely to be steer-days/acre.

The unit is based on an individual's experience of what it takes to maintain a herd of critters on a particular pasture for a day. This is not easily taught in a grazing school or learned from a book. It is strictly a learn-by-doing approach.

To keep this kind of record, think about a checkbook ledger. As a matter of fact, an actual check ledger can be used or a computer spreadsheet or a pocket notebook or just about anything. Here is the basic approach: Every two weeks, walk or ride all the paddocks and record your estimate of how many cow-days/acre (CDA) are available in each paddock for that date. Most recently grazed paddocks may be zero while most rested paddocks may be 40 or 50.

Tally up total CDAs available and compare that to the herd requirement for a two week period. If there are CDAs left over, think about cutting hay or adding more stock. If there are not enough CDAs to cover demand, adjust stocking or rotation frequency. Slowing down rotation to provide more rest will allow CDAs to accumulate. Comparing today's count to previous estimates allows you to determine whether growth rate is accelerating, maintaining, or slowing. Keeping an ongoing record of pasture use allows you to compare your estimates to what actually occurred. It can help fine tune your grazier's eye.

For those who want an actual measure of forage yield, there are both simple and complex approaches.

The most complex approach is the clip-and-weigh technique commonly used in research. Standing forage within an enclosure or quadrat of known size is clipped, allowed to air dry, and then weighed. A subsample can be dried in a microwave oven to more accurately determine actual dry matter yield. A constant multiplier term relative to quadrat size is then used to calculate yield per acre. The problem is that it takes 10 to 20 quadrats per paddock to get good data (85-90% accuracy) so most graziers lose interest in clipping samples pretty quickly.

Fortunately there are other methods, less accurate but simpler, including predicting yield from height, weighted disc, rising plate, and electronic capacitance meter. These methods are referred to as "non-destructive" because no forage is actually harvested.

For any of these methods to be reliable, someone, somewhere has done quadrat clipping in conjunction with one or more of these non-destructive techniques. The reliability of non-destructive yield estimates is directly proportional to how well the calibration was done, which basically means how many quadrats were clipped to develop the prediction equation. For those of you who wonder if algebra is ever used in the real world, the answer is yes and yield prediction is where it is used.

Beginning with the simplest technique of predicting yield from height, the taller the pasture is, the greater the yield, as a general rule. I can already hear some people getting bent out of shape in disagreement as they think about Kentucky bluegrass or perennial ryegrass in contrast to orchardgrass, so we need to think in context of pasture type.

The following table can be found in a number of grazing publications and on the Missouri Sward Stick. It was developed by measuring sward height and clipping tens of thousands of quadrats.

The stand condition accounts for pasture stand density with an excellent stand being greater than 90% ground cover, good being 75% to 90%, and fair 60% to 75%. When stand density is less than 60%, growth habits of the plants begin to change, and prediction of yield from height becomes unreliable.

Other states have developed similar relationships for other types of pastures. When properly done, predicting yield from height can be done with 80% to 90% accuracy

If a universal value of 300 lb/acre/inch is used, the reliability drops to 50% to 60% Estimated dry matter yield in pounds per acre-inch for several pasture types and stand conditions.

PASTURE SPECIES	STAND CONDITION		
	Fair	Good (lb/acre/inch)	Excellent
Tall Fescue + N	250-350	350-450	450-550
Tall Fescue + Legumes	200-300	300-400	400-500
Smooth Bromegrass + Legumes	150-250	250-350	350-450
Orchardgrass + Legumes	100-200	200-300	300-400
Bluegrass + White Clover	200-400	400-500	500-600
Mixed Pasture	200-300	300-400	400-500

Weighted disks and rising plates measure compressed sward height and account for variation in stand density by sinking or rising to different levels in response to sward resistance to compressing. A weighted disk is usually a sheet of plexiglass about two feet in diameter with a hole in the center with a measuring stick through the hole. The disk is dropped from a constant height and comes to rest wherever the sward stops it. If the disk has been properly calibrated, yield is then determined from the resting height of the disk.

A rising plate works very similarly except that the plate is rested on top of the sward and a measuring stick pushed through it to ground level.

A simple weighted disk can be constructed for less than $20 while a good rising plate can be purchased for $300 to $400. Properly calibrated, a rising plate provides 80% to 90% accuracy. Improperly calibrated, accuracy drops to 50% to 60%.

Electronic capacitance meters predict yield based on electrical charge of the forage relative to its moisture content. Within the range of normal moisture content of vegetative forage, capacitance meters are fairly reliable. Reliability declines with drought conditions, mature forage, or winter stockpile forage. As with any other non-destructive technique a calibration equation must be developed. With proper calibration and sampling technique, accuracy can be 80% to 90%. And, you guessed it, with improper calibration accuracy drops to 50% to 60%.

Units range in cost from $400 to over $1,000 depending on memory and computer interface functions.

Some people get the idea that because a rising plate or capacitance meter give you a specific number, such as 2867 lb/acre, that this is somehow the "right" answer. That value is only an estimate of the true yield and may be no nearer the "right" answer than your eyeball estimate.

The reliability of that estimate depends on consistency of the sampling technique and soundness of the prediction equation. A well trained eye can be more reliable than the $1000 instrument with improper technique and calibration.

Whichever technique is used, a half day every two weeks spent on measuring pasture availability can provide a wealth of information on forage supply, pasture growth rates, and rest period requirements. Once the information is in hand, a running record can be maintained either on computer or paper.

Just think of it as your check ledger. Whether it's the bank account or your pasture, failure to watch the balance can be both expensive and embarrassing.

The Basics:

- MiG allows the pasture manager to easily assess and keep a record of pasture condition and supply.
- Monitoring forage supply can be calculated by animals or by forages.
- The animal-based approach, everything is expressed on an animal unit or class of livestock basis with no actual estimate of forage dry matter yield ever considered.
- The most complex approach is the clip-and-weigh technique commonly used in research. Standing forage within an enclosure or quadrat of known size is clipped, allowed to air dry, and then weighed.
- Predicting yield from pasture height can provide reliable estimates of forage availability.

Think About Your Farm or Ranch:

- Every two weeks, walk or ride all the paddocks and record your estimate of how many cow-days/acre (CDA) are available in each paddock for that date. Tally up total CDAs available and compare to the herd requirement for a two week period. If there are CDAs left over, think about cutting hay or adding more stock. If there are not enough CDAs to cover demand, adjust stocking or rotation frequency.
- Keeping an ongoing record of pasture use allows you to compare your estimates to what actually occurred and can help fine tune your grazier's eye.

23 How to Judge Maximum Intake of Forage

The basic rule is the more you eat, the fatter you get, but that "more" has to mean "more total nutrients," not just "more bulk weight."

Now that we've discussed forage quality, it is time to shift our focus to getting that high quality forage into the critter. What the animal will consume of its own free will is what we in research circles like to call voluntary dry matter intake or VDMI. In daily conversation we usually just say "intake" and assume everyone knows what we are talking about.

Just so we are all on the same wavelength, in this discussion the bottom line on intake is going to be consumption of megacalories (mcal) of energy and pounds of digestible protein. We often talk about intake as pounds of forage dry matter or intake as a percent of bodyweight, but in the end, animal performance is determined by level of nutrient intake. The basic rule is the more you eat, the fatter you get, but that "more" has to mean "more total nutrients," not just "more bulk weight."

Intake of dietary nutrients is determined by the combination of dry matter consumption and the nutrient density of that dry matter. A steer consuming 12 pounds of a forage with .7 mcal/lb will gain better than a steer consuming 15 pounds of forage with .5 mcal/lb. So even though we got the second steer to eat a higher percent of bodyweight and more total pounds of forage, it didn't gain as well because net energy consumption was lower.

A lot of my nutritionist friends would likely jump all over the above example, so I will beat them to the punch. What is wrong with the previous picture is that we all know that a grazing animal is going to be able to consume more of the higher quality forage than the lower quality forage because of a little phenomenon known as rate of passage.

There are several factors that can regulate intake and one of the internal mechanisms in the rumen is that bulk forage can only be pushed through so fast. The example forage that had .5 mcal in all likelihood has a higher fiber content than the .7 mcal forage. The fiber component that we believe regulates rate of passage and subsequent intake is neutral detergent fiber or NDF.

More specifically, the indigestible component of NDF appears to be a primary regulator of intake on pasture. On low quality pastures, dietary fiber is the main limitation on intake. So what about high quality pasture?

Another way to think about intake regulation from the animal's perspective are the three grazing components of daily intake: time spent grazing, biting rate, and bite size. Ruminants are pretty well unionized and will only work so many hours a day. Cattle roughly divide their day into three nearly equal parts. They usually spend about eight hours grazing and eight hours ruminating and, because these are strenuous activities, they need to rest about eight hours. Which of these things can we control to enhance intake?

Under ideal pasture conditions, most ruminants can reach maximum required intake in about four hours. In poor conditions they may extend their grazing time to ten hours, but that is just about the biological max they can sustain. We can make pasture conditions as near to ideal as possible and reduce the amount of time they spend grazing, but we cannot expect them to compensate for our poor management by spending more time grazing.

Biting rate is genetically regulated. There are some breeding lines which take more bites per minute than others, but the differences are not large. Sheep take many more bites per minute and per day than cattle. Cattle may take around 24 to 30,000 bites per day while sheep can exceed 50,000. That is more than most teenagers. If we create poor pasture conditions, we can not expect our grazers to take more bites per day to compensate.

The one factor that we can really control is size of the bite. If each mouthful the animal takes is large, relatively speaking, less time is required per day for grazing and biting rate will slow a bit as more time is required for chewing and swallowing the larger bite. Pastures that are dense and of optimal height provide the best biting opportunities.

Fine leafed grasses usually are consumed at a higher rate than coarse grasses because each cubic inch of the canopy contains more forage mass. Thus, each bite contains more mass.

Height is another critical factor, particularly for cattle because of their tongue use in grazing. Cattle achieve large bites by wrapping the tongue around forage and sweeping it into the jaws where it is sheared off.

If pasture height drops below three or four inches, it becomes very difficult for cattle to take very large bites. Think about the difference in mouth structure and components of sheep and cattle and you quickly start to understand the range wars of the 19th century. Sheep can thrive on pasture that a cow would starve on.

If pastures become too tall, grazing efficiency is reduced due to declining leaf mass in the upper part of the canopy and change in quality with each deeper bite into the canopy. There is an optimal range to maintain pastures to ensure the desired intake level. We generally talk in terms of six to ten inch turn-in height and 3 to 4 inch residual, but it depends on species, time of year, animal demand, and all those other weasel factors we always throw out there to cover our fannies.

One of the rules of thumb I use for cattle grazing to try to make sure intake is adequate is what I call the "eyes and nostril rule." If cattle go into a new paddock, heads down grazing, and you can't see their eyes, the pasture is probably too tall or mature and quality will restrict intake. If you can see their nostrils, the pasture is too short and availability will limit intake. So the guideline for maintaining intake is to try to keep forage availability between the bottom of the eyes and top of the nostrils.

The Basics:

- Intake of dietary nutrients is determined by the combination of dry matter consumption and the nutrient density of that dry matter.
- On low quality pastures, dietary fiber limits intake.
- There are three grazing components of daily intake: time spent grazing, biting rate, and bite size.
- If we create poor pasture conditions, we can not expect our grazers to take more bites per day to compensate.
- Pastures which are dense and of optimal height provide the best biting opportunities.
- Fine leafed grasses usually are consumed at a higher rate than coarse grasses because each cubic inch of the canopy contains more forage mass.
- If pasture height drops below three or four inches, it becomes very difficult for cattle to take very large bites.
- If pastures become too tall, grazing efficiency is reduced due to declining leaf mass in the upper part of the canopy and change in quality with each deeper bite into the canopy.
- Try to achieve the "eyes and nostril rule" to maximize grazing.

Think About Your Farm or Ranch:

- How can you improve grazing conditions in your pastures?

24 Residual Affects Performance of Animals & Pasture

Residual consists of both green and dead plant material, but we are usually more concerned with the green material needed to create new pasture.

I don't know how many times I've heard a farmer or rancher say, "We move 'em when the paddock's grazed down all the way." That is a pretty casual remark to describe one of the most important day-to-day decisions a grazier makes. Residual is the term we use to describe the plant material left behind after grazing. The amount of post-grazing residual has a big effect on both pasture and animal performance.

We generally refer to residual as either sward height or forage mass. For instance we could have a "four inch residual" or a "1500 pound/acre residual."

Residual consists of both green and dead plant material, but we are usually more concerned with the green material as this is the photosynthetic factory that is needed to create new pasture. Some species, such as alfalfa, have large reserves of available energy as soluble carbohydrates stored in the roots. All perennial species have some stored energy whether in roots, rhizomes, or stem bases; but for most, residual leaf area is what determines regrowth rate.

For most forage species, regrowth rate is directly proportional to the amount of green leaf area remaining after grazing. Leaf area index or LAI is the term used to describe the amount of leaf area relative to ground area. An LAI of 2 means there are 2 square feet of leaf area per square foot of soil surface. While a pasture with a residual LAI of 2 does not necessarily regrow at twice the rate of a pasture with LAI of 1, it will recover from grazing at a much faster rate. This is particularly true with species such as smooth bromegrass or indiangrass which do not maintain much leaf area close to the ground.

Grasses like Kentucky bluegrass, tall fescue, and perennial ryegrass may have an LAI greater than one even when grazed down to only an inch or so. The residual may appear to be very short, but they may have more leaf area in that short height than smooth bromegrass has in 3 or 4 inches.

This is why one-height-fits-all grazing recommendations won't work in diverse pastures.

Repeatedly grazing to short residuals will drive the pasture towards the species with the highest leaf area below the physical limitation of grazing. White clover and kura clover are two legumes that fall into the category of high residual leaf area below normal grazing heights. Other taller growing legumes like alfalfa and red clover rely more on stored carbohydrates for regrowth.

Pastures that persist with very short residual are not necessarily undesirable. Kentucky bluegrass-white clover is one of the highest quality pastures found in the eastern half of the USA. Perennial ryegrass-kura clover would be a fantastic pasture for dairy cows or any class of livestock with high energy and protein requirements. The drawback in temperate climates is that pastures grazed in this manner tend to be shallow rooted and very prone to drought stress. If ample soil water is not always readily available, regrowth rate becomes extremely slow and we quickly run out of paddocks.

This was a lesson we learned at FSRC in the drought of 1988-89. Short grazed paddocks required 50 to 60 day rest

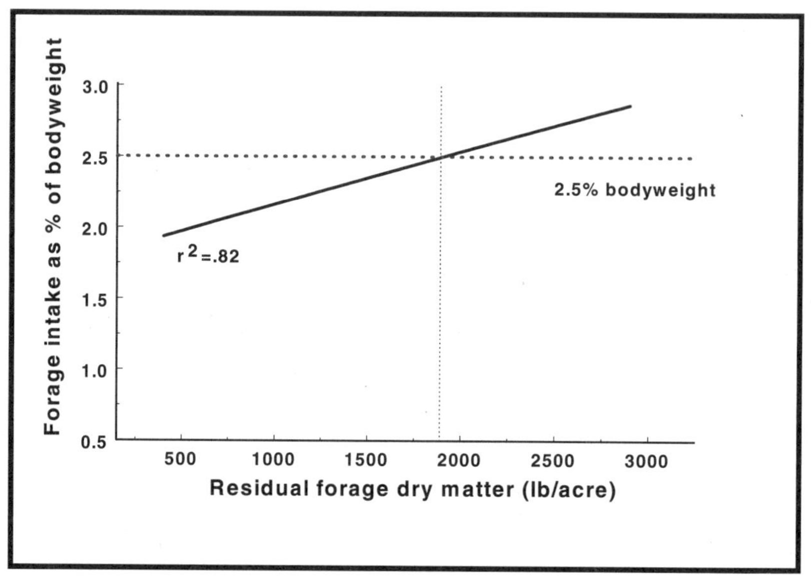

periods to grow back to the same forage mass that paddocks left with greater residual were achieving in 35 to 40 days. During the 1999-2000 drought, we were never short on pasture, largely because we had learned how to manage residual forage to maintain growth rates and forage availability.

Taller residuals tend to favor taller growing species which also tend to be deeper rooted species. During drought periods, the deeper rooted species will usually regrow at a faster rate than shallow rooted plants. Varying the grazing residual through the season can shift the balance of species from shorter to taller species.

To maintain a wide range of diversity in pastures, plan to do just that, vary the residual. To be successful, you'll need to know what species are present in the pasture and how they will respond to changing grazing height.

From the animal's perspective residual height or forage mass is what limits bite size and, subsequently, forage intake (See above figure). How significant that limitation becomes depends a lot on the length of the grazing period. Animals grazing for shorter periods can tolerate shorter residuals than animals grazing for longer periods. We will discuss this in detail in the next chapter.

Another important consideration is how forage quality changes through the pasture canopy. The youngest plant parts, which are usually the highest in the canopy, are the most nutritious. Leaving taller residuals ensures within a single grazing period that forage intake will be higher quality.

Grazing shorter in a single grazing period will result in lower quality forage being consumed. But the paradox is that repeatedly grazing to short residuals generates higher quality forage in subsequent grazing periods while leaving taller residuals provides for lower quality forage for future grazing.

The appropriate residual is also going to depend somewhat on the class of livestock being grazed and what the performance targets might be. Lower milk producing cows can be grazed to lower forage residuals than high milk producing cows because of their lower energy demand.

Dry cows at maintenance are the class of livestock that we push the hardest to clean up pastures and, subsequently, leave the lowest residuals. Sheep and horses by virtue of their ability to bite grass shorter will typically graze pasture to shorter residuals than will cattle and still maintain adequate intake. When grazing mixed livestock species in a rotational system, decisions of when to move livestock should be based on the animal class requiring the highest residual, not those requiring the lowest.

Besides just plant and animal performance, there are other factors in the production system which are also affected by the amount of residual we leave after each grazing. Three that come quickly to mind are increase or decrease in soil organic matter, water infiltration and runoff processes, and plant litter and manure decomposition cycles. We will look at each of these factors in more detail in future issues.

So the big question ends up being, what is the appropriate residual? As usual the answer is it depends. From the pasture perspective it depends on the forage species present, what our goals for composition of the pasture are, growing conditions, and time of the year.

From the animal perspective, we need to consider performance targets, class of livestock, and stage in the production cycle with gestating and lactating females.

There is no substitute for understanding plant, animal, and soil needs when making decisions about how much residual to leave behind.

The Basics:

- One of the most important decisions a grazier makes each day is when to move animals to the next paddock.
- All perennial species have some stored energy whether in roots, rhizomes, or stem bases; but for most, residual leaf area is what determines regrowth rate.
- Regrowth rate is directly proportional to the amount of green leaf area remaining after grazing.
- Repeatedly grazing to short residuals will drive the pasture towards the species with the highest leaf area below the physical limitation of grazing.
- Taller residuals tend to favor taller growing species which also tend to be deeper rooted species.
- During drought periods, the deeper rooted species will usually regrow at a faster rate than shallow rooted plants.
- Varying the grazing residual through the season can shift the balance of species from shorter to taller species.
- Repeatedly grazing to short residuals generates higher quality forage in subsequent grazing periods while leaving taller residuals provides for lower quality forage for future grazing.
- When grazing mixed livestock species in a rotational system, decisions of when to move livestock should be based on the animal class requiring the highest residual, not those requiring the lowest.

Think About Your Farm or Ranch

- List the plant species in your pasture. How does each respond to changing grazing heights?
- The appropriate residual is also going to depend somewhat on the class of livestock you're grazing and what your performance targets are. What are your goals for pasture diversity, growing conditions, and seasons?
- What are your animal performance targets, class of livestock and stage of their production cycle?

25 Length of Grazing Period: Does It Really Matter?

For the serious grazier,
the answer needs to be based on specific objectives
for the pasture and livestock.

How often do you want to shift your livestock to a new paddock? While for some people the answer might be based on their work ethic or basic laziness, for the serious grazier, the answer needs to be based on specific objectives for the pasture and livestock. Let's look at a few possible scenarios and think about what is happening at the plant-animal-soil interface.

Our first visit is going to be a beef cow pasture where the owner works off the farm and mostly has the cows so he can justify driving his dually pickup to work every day. His weekends are spent fixing broken fence along the highway and baling little dabs of hay around the neighborhood so he has some winter feed. He knows he'll need it because he is always out of grass by November. Because he's busy, he doesn't have time to move cows but once or twice over the entire summer.

As we walk out across the pasture we see it is mostly tall fescue, Kentucky bluegrass, and white clover. There is a bit of crabgrass around and a few thistles. Patch grazing is very apparent, with bluegrass spots chewed down nearly to the dirt and fescue clumps boot-top high. The kindergarten kids could come out and not worry about stepping in fresh manure because the piles are few and far between.

We are seeing the typical effects of unrestricted grazing. How much of these effects are attributable to length of the grazing period? All of them.

Just up the road we'll stop in at another cow-calf operation. This fellow also works in town but rather than spending the weekend making hay and fixing worn out fence, he spent his time building some new fence and laid some water line and found he can move a hundred cows to a new pasture in 15 minutes on his way to work. He'll move them two, or maybe even three times a week.

As we walk across this pasture, we see a wider array of plants, including a lot more legumes and even some native bluestem showing up in some of the pastures. There is a lot less spot grazing, although some overgrazed spots are still apparent. There are visibly more fresh manure spots around but we see they tend to be more scattered plops and strings rather than big piles.

How much of these effects are attributable to length of the grazing period? Most of them, but now length of the rest period has also come into play.

The third farm we'll visit is a grass-based dairy. Here the owner takes his grass pretty seriously and doesn't work in town. The cows get moved to a new paddock after every milking, so the length of the grazing period is less than 12 hours. We note the pasture is grazed much more evenly, although there are some high spots and low spots. We see little exposed soil. Most of the surface is covered with green, growing plant material representing a wide range of plant species. You can hardly take a step without getting green, gooey pooh on your shoes.

This guy used to move his cows every couple of days, but he found when he started moving them after every milking, his tank average jumped about seven pounds.

How much of these effects are due to length of the

grazing period? Most of them, but attention to management detail has come into play.

In the last chapter we discussed the importance of grazing residual. That is what is left behind after grazing. As we look at length of the grazing period, it is critical we keep in mind the importance of leaving an appropriate amount of residual. A shorter grazing period does not automatically produce positive benefits for the pasture. A shorter grazing period leaving proper residual will usually improve a pasture quicker than a longer grazing period leaving the same residual.

So what is the force that makes shorter grazing periods have such an impact on pasture? Let's assume we are going to hold stocking rate constant, that is, we will keep the same number of animals on the farm. If we have 100 cows on 200 acres, the stocking rate is one cow per two acres. If our cows weigh 1000 pounds apiece, stocking rate can also be expressed as 500 pounds liveweight per acre. Now we need to consider stock density.

While stocking rate describes the animal pressure applied to the farm on an average or seasonal basis, stock density describes animal pressure in the immediate sense or: how many animals on this acre on this day.

On the first farm we visited stocking rate and stock density were the same because the animals were never concentrated.

On the second farm, stock density was increased because animals were confined to only a part of the pasture. For a two to three day rotation frequency to work effectively, 20 paddocks would likely be required.

Given 200 acres divided into 20 paddocks, each paddock is 10 acres and with all 100 cows on it, the stocking rate is still 500 lb liveweight/acre, but the stock density is 10,000 lb liveweight/acre. The animal pressure on the plant community has just increased 20-fold.

On the dairy farm, we will likely have at least 50 paddocks, so now we have 100 cows on 4 acres and the stock density is 25,000 lb liveweight/acre or 50 times the animal impact of the first farm.

Stock density is the tool graziers use to create desirable pastures, manage the nutrient cycle, and control dietary intake. Shortening the grazing period is the most feasible way of increasing stock density on almost any farm or ranch.

When a pasture is grazed at high stock density, the likelihood of any plant being grazed is increased because there are more mouths per acre, more bites being taken on every acre, and the grazing animals become more aggressive in their grazing behavior. The grazing manager can decide to move the stock off the paddock at any time. The paddock can be grazed to any target residual the manager desires, as long as the grazing is being monitored.

As a general rule, the higher the stock density, the more uniform the grazing. Managing grazing residual and uniformity can help manipulate the plant community to meet your specific goals.

From the animal perspective, length of the grazing period has a great deal of effect on forage intake. Previously, the importance of forage residual on forage intake has been emphasized. Usually, shorter residual means lower intake. But what about grazing to the same target residual, but getting there in different lengths of time?

Let's look first at the twelve-hour grazing period on the dairy farm. When the cows graze the pasture to a 1200 lb residual in twelve hours, the bulk forage material reaching the rumen is a mixture of leaves and stems, green and dead, basically the whole plant. The high quality cell contents in the leaves allow the rumen bugs to do a good job of digesting the lower quality stem material. Now let's take a week to graze the pasture to the same 1200 lb target residual.

On the first couple of days of the grazing period, intake is superb as all that the cows are eating are the leaf tips. Very little stem or dead material is entering the rumen. By the middle of the week, they are eating more stem material, but there is

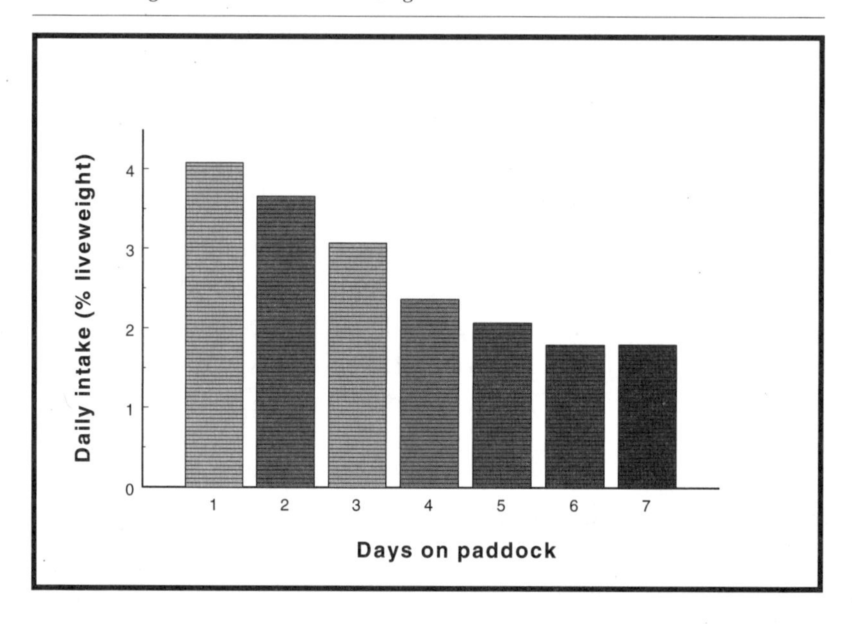

less of the cell content goodies going into the rumen to help digest the lower quality material. By the end of the week, there is very little leaf material going in but a whole lot of stems. The end result is a 50% reduction in intake by the end of the week. Figure 5 illustrates the daily change in forage intake as we actually measured it in a FSRC grazing study with big bluestem pasture.

We have conducted several studies at FSRC and repeatedly found this trend to occur. The bottom line is if you want to maintain a target level of intake to achieve a particular milk production or weight gain target, for each additional day the animals remain on a paddock, a greater forage residual must be left.

Most of you probably had to read that sentence twice. Common sense tells you that you can't do that.

How can you leave more grass in the paddock each additional day the animals are in the paddock? You can't.

At any given stock density, you either shorten the grazing period or sacrifice performance. This is the grazier's paradox.

Do you graze to achieve a target forage residual or do you graze to achieve a target level of animal performance?

Can you achieve both? Only if the pasture and stock targets are compatible.

And lastly, what about those manure piles? Common sense tells us that if more animals are standing around on a paddock, there will be more manure piles on the ground when they leave. The concentrated effect of those piles and the urine you can't see is a pulsing of nutrients into the system.

On the first farm we saw spot effects of where those few piles had fallen but on the third farm we saw a flush of nutrient-stimulated plant growth. The shorter the grazing period, the more efficient the natural nutrient cycle becomes. If the residual is being managed for optimal intake of high quality forage, the piles change to scattered plops or lakes of green goo rather than piles of brown wood chips.

Length of the grazing period does matter and it matters a great deal.

The Basics:

- A shorter grazing period leaving proper residual will usually improve a pasture quicker than a longer grazing period leaving the same residual.
- While stocking rate describes the animal pressure applied to the farm on an average or seasonal basis, stock density describes animal pressure in the immediate sense or: how many animals on this acre on this day.
- Stock density is the tool graziers use to create desirable pastures, manage the nutrient cycle, and control dietary intake.
- Shortening the grazing period is the most feasible way of increasing stock density.
- The higher the stock density, the more uniform the grazing.
- To maintain a target level of intake to achieve a particular

milk production or weight gain target, each additional day the animals remain on a paddock, a greater forage residual must be left, yet at any given stock density, you either shorten the grazing period or sacrifice performance. This is the grazier's paradox.

- The shorter the grazing period, the more efficient the natural nutrient cycle becomes.

Think About Your Farm or Ranch:

- How often do you want to shift your livestock to a new paddock?
- How can you manage grazing residual and uniformity to help manipulate your pastures' plant community and meet your specific goals?
- Will you graze to achieve a target forage residual or will you graze to achieve a target level of animal performance? Can you achieve both?

26 Everybody Needs a Little Rest and Recuperation

Pastures will be more productive and stable if allowed to rest and regrow.

When you've been working hard, you're always looking forward to a little R&R — rest and recreation in the people world, but rest and recovery in the plant world.

Why do businesses provide employees with paid vacation? If you think it is because your employers are really wonderful people who care about you as an individual human being, guess again.

It is probably because all the studies on worker productivity show that employees are more productive if allowed some R&R occasionally. The same concept applies to your pastures. They will be more productive and stable if allowed a little R&R.

Which brings up the big question. How much rest and how often is it required?

If we look across a wide range of businesses, we find all sorts of work schedules. There is the standard 40 hour week with five 9 to 5 days followed by the standard two-day weekend off or we might opt for four 10-hour days with three days off. In some lines of work we might find 10 days on and four days off or even 20 days on and ten days off. It is quite apparent that there are a lot more ways to get the job done other than the standard American 5-day work week. What all of these work options try to accomplish is increasing worker productivity by finding the optimum balance of working time and R&R.

We can draw some more parallels between our working world and pasture management. Where do most of the standard 9 to 5 days occur? This is the norm in what I would call a pretty cushy environment. The worker isn't pushed real hard and so the R&R requirement is not real great.

Where do we find 20 day shifts with longer R&R? Off-shore oil drilling, back country mining, and logging would be good examples. Harsher environments where the work may be more physically exhausting, but it is expensive to get workers in and out of the job site.

The bottom line is rest can be fairly minimal in soft environments but had better be more extended in hard environments.

Have you ever noticed that when the weather is pleasant and things are going well, you might not feel like you need that extra day off? But when the weather gets a little nasty, whether that be too hot or too cold, or rainy and overcast, you feel like you need a break from work a little more often, otherwise you get depressed or snappy?

Draw the parallel with pasture again and we find that pasture in fine weather needs a lot less rest than pasture in foul weather. Now, a pasture's idea of foul weather may be a little different than yours, but it is surprisingly similar regarding temperature. The one big difference is that pastures tend to like those damp, rainy days a lot more than we do.

Rest is critical for maintaining pasture vigor and productivity. When a pasture is rested several things occur. The first and foremost is that the plant has the opportunity to grow more leaves. Leaves are the critical component of the photosynthetic process, which provides plants with energy to live.

From energy generated in photosynthesis, the plant may grow more leaves, stems, and roots, or may deposit energy as stored carbohydrates (CHO) in various parts of the plant. While

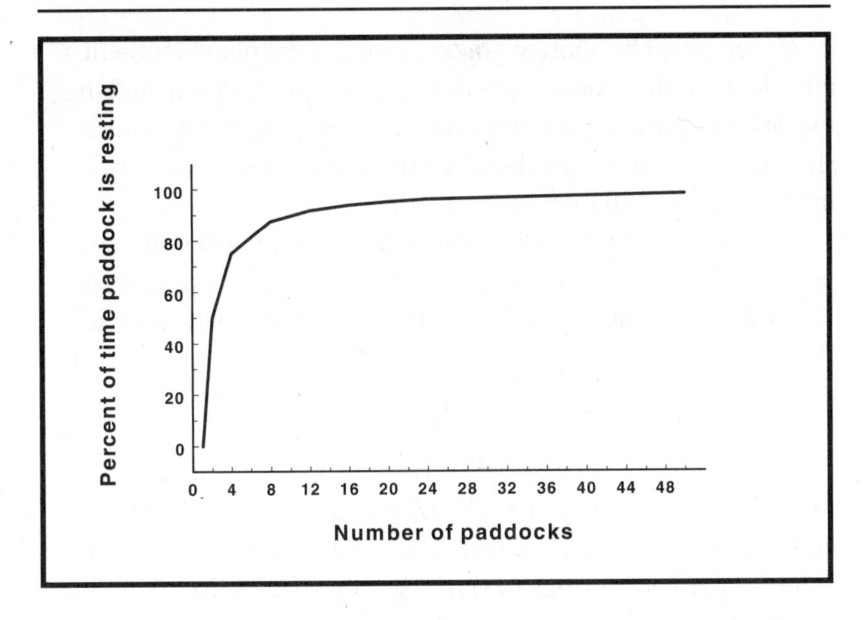

we generally think about storage CHOs in the roots, they may also be stored in stem bases, rhizomes, or stolons. The basis for the length of the rest period is how much time is required for the plant to reach a positive CHO balance.

That is, when does the plant start restoring reserves rather than drawing on them? This depends very much on the particular plant or type of plant in question, growing conditions, extent of defoliation, and CHO status before grazing occurred.

With grazing residual, discussed earlier, the difference among plant species is the amount of leaf area remaining below grazing height. In general, species with high leaf area close to the ground require less rest that species with minimal leaf area in the residual.

Plants with high leaf area below grazing height rely more on active photosynthesis to maintain energy level than low leaf area plants. A general rule here is grasses typically require shorter rest periods than legumes. So how do you manage a mixture of the two?

Because legumes usually rely on stored CHO for re-growth, they can be grazed shorter than grasses and still recover rapidly. The shorter grazing reduces the photosynthetic advantage of the grass by restricting available leaf area and the two species grow back fairly uniformly. Repeatedly grazing to short residual favors legumes in a rest period while leaving taller residuals will favor grasses.

When growing conditions are favorable, meaning near optimum temperatures, adequate soil moisture, and appropriate fertility, photosynthesis is very efficient and regrowth is rapid. As any one or more of the growing conditions shifts away from optimum, the rest period must be extended for the plant to regrow to a comparable CHO level.

In regions where both warm-season and cool-season forages are grown together, changes in growing temperature and soil moisture can rapidly shift the balance of power away from cool-season species to warm-season species. Maintaining a cool-season and warm-season balance in mixed pastures is one of the most challenging tasks for a grazier requiring frequent adjustments to rest period timing and duration.

How the pasture has been treated earlier in the season or even in previous years, particularly in brittle environments, also has a profound effect on rest period requirement. If a pasture has been used lightly early in the season, it can be grazed with a shorter summer rest period without unduly affecting vigor or production.

If a similar pasture had been used heavily through the early months of the grazing season, using a short summer rest period could severely reduce the vigor of the pasture. The difference being CHO status prior to grazing. Even if grazed to identical residual, the pasture having higher storage CHO will grow back more rapidly.

So next time you're kicked back with your feet in the air and a cold one in your hand, make sure your pastures are getting their fair share of R&R also. It makes the rest of the work week a lot more endurable.

The Basics:

- Pasture in fine weather needs a lot less rest than pasture in foul weather.
- When a pasture is rested the plant has the opportunity to grow more leaves. Leaves increase the photosynthetic process, which provides plants with energy to live.
- Species with high leaf area close to the ground require less rest that species with minimal leaf area in the residual.
- Grasses typically require shorter rest periods than legumes.
- Repeatedly grazing to short residual favors legumes in a rest period while leaving taller residuals will favor grasses.
- As any one or more of the growing conditions shifts away from optimum, the rest period must be extended for the plant to regrow to a comparable carbohydrate level.
- Maintaining a cool-season and warm-season balance in mixed pastures is one of the most challenging tasks for a grazier requiring frequent adjustments to rest period timing and duration.
- How the pasture has been treated earlier in the season or even in previous years, particularly in brittle environments, also has a profound effect on rest period requirement.

Think About Your Farm or Ranch:

- How much rest are you giving your pastures? Why?
- Can changing your rest pattern change your pasture composition to something more desirable?

27 Old McDonald Had a Farm with Many Species

The variance that occurs in plant communities offers potential for grazing by a number of different livestock and wildlife species.

It is funny how some childish thoughts make so much sense but are so difficult for some adults to grasp.

We have this old farmer who has a remarkably wide array of critters on his place. While the kids think it is really cool, we adults have to shake our heads and question his business acumen for not specializing in something he is good at. As usual, the truth probably lies somewhere in the middle.

Specialization of a single enterprise or type of livestock on a farm or ranch does have very real advantages, as most of the conventional agriculture press keeps reminding us. But specialization also has disadvantages which diversification can help overcome.

Specialization is especially valuable and easy when done in confinement where environmental conditions can be controlled. As soon as we step outdoors and experience both environmental and landscape variations, the value of diversification of livestock enterprises becomes much more apparent.

The variance that occurs in plant communities offers potential for grazing by a number of different livestock and wildlife species.

One of the tools that can be used to capitalize on the diverse species composition of pasture and rangeland is mixed species grazing or the use of different types of livestock in the same grazing unit.

To manage vegetation we need to first understand foraging preferences of different species of animals. One very important concept is that of complementarity versus competition.

If two species have exactly the same grazing preferences and habits, there is no biological reason for adding the second species. There may be economic reasons for mixed species grazing such as counter-cyclicity of prices, multiple marketing options, and others which can help the bottom line.

We are usually looking for two species having different diet preferences and foraging habits so that the species complement one another for optimizing forage and landscape use.

A single enterprise farm or ranch develops a very set pattern of resource usage which moves the resource base in one, fairly predictable direction.

For example, a stocker operation in a range environment may only have cattle on the ranch from April through July and it likely gets used pretty heavily during that time period. The plant community will likely decrease in the species most preferred for grazing in spring and early summer while increasing in species, that are later maturing or less preferred for grazing.

Repeating the same pattern of use year after year will slowly but steadily move the plant community in the direction of late maturing, less preferred plants. Over time the plant community becomes less and less adapted to the use being made of it.

A system like this would benefit from periodically applying late season use while allowing rest in the spring.

Alternatively, overlaying the cattle operation with another stock enterprise to apply grazing pressure to the less preferred pasture species would take away some of their competitive advantage when cattle were removed from the pasture. As grazing managers, it is our responsibility to keep the pasture or range in the desirable condition for productive grazing.

Even in season-long grazing systems there are species of plants that cattle do not like to graze but may readily be grazed by other foraging species. Leafy spurge being a good example.

One of the few bright spots on the invasive plants front is the use of sheep to control leafy spurge, as well as a few other undesirable exotic species. The greatest challenge is that the leafy spurge occupies more acres than our demand for lamb and wool can possibly accommodate.

The most commonly used mixed species grazing system worldwide is probably the combination of cattle and sheep. Sources vary in the exact numbers, but cattle are generally considered to consume about 80% to 90% of their diet as grass with forbs and browse making up the balance in nearly equal parts. Sheep will consume 50% to 60% grass and 40% to 50% forbs and browse with about a 2 to 1 ratio of forbs to browse.

If you operate in a forage environment high in forbs, sheep would be a very good addition to a cattle operation to increase meat production. Because sheep are consuming something different from the cattle, more of the sunlight captured on every acre is being converted to a salable product.

Bison, on the other hand, have an even stronger grass orientation than do cattle. Adding bison to the same operation would place the cattle and bison in a competitive situation rather than complementary.

Goats have a much higher browsing preference than either cattle or sheep. Typical figures indicate 60% - 70% browsing on woody species by goats with the remaining 30% to 40% split evenly between grass and forbs. So if your problem is woody species invasion in cattle pastures, goats would be a better choice for vegetation management than sheep.

For the large wildlife species commonly found on grasslands, elk and bison are primarily grass grazers while deer and antelope are browsers.

So far we have only talked about the biological or ecological aspect of mixed species grazing. In either the sheep or bison example there may be economic benefit from adding one of these species to the cattle operation. Thus, both biological and economic considerations must be weighed.

A biologically complementary species may have both positive and/or negative economic consequences.

Adding sheep to the system may increase net meat production per acre and provide another payday during the year, but if additional fencing or modification of the cattle watering system is required, the additional income may not offset the additional costs.

If sheep control a serious weed problem that the cattle were not controlling and pasture production was declining, there may be an increase in cattle production that is directly attributable to the presence of sheep in the system after the weed is brought under control.

If adding sheep eliminated the need for clipping or spraying, that would be a factor assigned to benefits in the sheep enterprise partial budget.

On the Gerrish farm we started out with just sheep and added cattle about four years later. One piece of property we bought had a terrible iron weed infestation. After several years of sheep grazing we saw very little ironweed. However in 1994 we liquidated the sheep operation after nine years.

Two years later, with just cattle grazing, that pasture was back to being covered with ironweed. We had to mow the pasture every year for about four years to bring it back under control. Sometimes we don't realize how important a particular part of the grazing operation is until we change it.

I am still not sure why Old McDonald had so many different types of livestock on his farm, but he appears to have either been a pioneer of horizontal integration in industrialized agriculture or a practitioner of mixed species grazing. From the pictures in the book, I think mixed species grazing was his objective because the cows and pigs and sheep and ducks and goats and chicks are all in the same pasture.

And note how green and lush his pasture always looks.

The Basics:

- To manage vegetation we need to first understand foraging preferences of different species of animals. If two species have exactly the same grazing preferences and habits, there is no biological reason for adding the second species.
- If you operate in a forage environment high in forbs, sheep would be a very good addition to a cattle operation to increase meat production.
- If your problem is woody species invasion in cattle pastures, goats would be a better choice for vegetation management than sheep.
- For the large wildlife species commonly found on grasslands, elk and bison are primarily grass grazers while deer and antelope are browsers.

Think About Your Farm or Ranch:

- What environmental problems or factors on your farm can be solved or enhanced by the addition of more species?
- What are the economic considerations for adding more species to your operation?

28 Follow the Leader For Grazing Efficiency

The basic objective of leader/follower grazing is to increase the biological efficiency of the pasture resource.

A leader/follower grazing system is one in which two or more classes of livestock having distinctly different nutritional needs or grazing habits are grazed successively in a pasture. The animals with higher requirements are allowed to selectively graze to ensure high individual performance, while the follow-up group with lower nutritional requirements are forced to clean up the less desirable plant species and plant parts. This cleanup grazing also ensures the sward will be more uniform, high quality forage for the next grazing cycle.

The basic objective of leader/follower grazing is to increase the biological efficiency of the pasture resource by more fully utilizing all the products of photosynthesis. Remember that not all products of photosynthesis are necessarily desirable. Weeds, for example.

Increased pasture efficiency may come in the form of increased animal output, reduction of undesirable plants, or by extending the useful grazing season.

Many producers initially fail to see how this approach can work in their program or don't feel they have the resources to implement a leader/follower system.

What does it take to successfully manage a leader/follower system? First, from an animal perspective, there must be two or more classes of livestock. In a cow-calf operation, this can be as simple as allowing calves to forward creep graze ahead of the cows. As long as the calves have access to higher quality forage before the cows do, it is leader/follower grazing.

Grazing your higher producing cows ahead of the lower producing cows is another option. The high producing cow will respond more to the improved forage quality allowed by selectively grazing than a lower producing cow. Grazing first calf heifers ahead of the mature cows is a similar option.

Combining a stocker and cow operation will often give better stocker performance than grazing stockers alone. At the Forage Systems Research Center, we have seen stocker gains .2 to .4 lb/hd/day greater with leader/follower systems compared to stockers grazing alone.

Other combinations are only limited by imagination.

In dairy systems, it is very common practice to graze dry cows behind the lactating herd. The primary goal is to better match available forage to animal nutritional requirements. Some dairies also split the lactating herd into high and low producing herds. This split can be based on either genetic ability or stage of lactation or both.

When sheep or goats are added to the equation, management can become more challenging but opportunities also increase significantly. Many producers view the sheep or goats as little more than brush beaters, but they are additional marketing opportunities in a diversified production system.

If weed or brush control is the prime objective, remember sheep are a little more efficient at controlling broad leaf herbaceous plants while goats are superior for control of woody brush. The challenge that come with adding sheep or goats is additional fencing is required compared to grazing just cattle, and water tanks sometimes need modification to let the shorter creatures access water.

Three-strand electrified hi-tensile works well for sheep and goat control in most cases. Once sheep or goats are well broke to electric fence, temporary fencing such as polywire or polytape are equally effective.

From the pasture standpoint, it is necessary to have a series of pastures for the two or more sets of livestock to rotate through. Each pasture should have individual access to water to keep the herds separate. To do an appropriate job of grazing utilization while maintaining adequate rest periods, a minimum of 10 to 12 pastures should be available, while more paddocks greatly increases management flexibility.

With fewer paddocks, it is difficult to keep the nutritional level high enough for the first grazers without shortening the rest period. Single strand electric fencing is generally adequate to keep the herds separate unless two breeding groups such as heifers followed by cows, are being utilized.

In the latter case, either multi-wire fences or keeping an open paddock between the herds is advisable. We have run up to three breeding herds in the grazing cell and just kept an open paddock between each herd throughout the breeding season.

To illustrate on farm use of leader/follower systems, I will describe the approach we often used in the fall of the year when we had both a cow-calf and ewe operation on our farm.

We typically weaned calves in late October or early November. This would also be when we were transitioning from active growth pastures to stockpiled pasture for winter grazing.

Weaned calves would be the first herd to hit stockpiled pastures. Our goal was to have the weaned calves graze no more than 25% of the available forage, usually in a two or three day grazing period. When the calves were moved ahead, the just dried off cows would come behind.

We did usually keep one empty paddock between cows and calves at this time.

We would allow the cows to harvest 30% to 40% of the forage. What was left behind after the first two groups was the forage around manure piles, forage that was under the brush, and residual white clover too short for cattle to harvest. These leftovers actually were the flushing pastures for our ewes.

We began breeding ewes about November 15 for April 15 lambs. By breeding late in the season, ewe fertility was naturally high.

The forage that was rejected by the cattle was rejected, not due to low quality, but due to physical location. Because sheep are not bothered by cattle manure, they would graze close to the dung piles, they would root forage from under the multiflora roses, and they could pick the little white clover close to the ground.

Without grain flushing, we typically marketed about 180% lamb crop from leftover pasture.

In the spring of the year, ewes with lambs would sometimes become the lead grazers due to the high percentage of twins we always carried. Due to changing production cycles and seasons you cannot always plan to use a particular species or class of livestock in the same sequence.

The bottom line is to match animal requirements with what the pasture can supply to make leader/follower grazing work efficiently.

The Basics:

- The primary goal is to better match available forage to animal nutritional requirements.
- You cannot always plan to use a particular species or class of livestock in the same sequence.
- Combining a stocker and cow operation will often give better stocker performance than grazing stockers alone.
- If weed or brush control is the prime objective, remember that sheep are a little more efficient at controlling broad leaf herbaceous plants while goats are superior for control of woody brush.

- Single strand electric fencing is generally adequate to keep the herds separate unless two breeding groups such as heifers followed by cows, are being utilized.

Think About Your Farm or Ranch:

- Do you have the resources to implement a leader/follower grazing system?
- What two classes of animals would work best in your system?

29 Managing Livestock by Creep Grazing

The key to creep grazing is the opportunity for offspring to select the choice forage first.

While "creep grazing" may sound like Leisure Suit Larry and the Lounge Lizards at a buffet social, it is actually a fairly old livestock management practice. Creep grazing is allowing the offspring to graze in a different location from its dam. Last chapter we talked about leader/follower grazing where a class of livestock with higher nutritional requirements is allowed to graze a paddock first and then the residual is cleaned up by a second group of animals with lower nutritional requirements. Creep grazing is a very specific type of leader/follower system.

There are two basic approaches to managing for creep grazing. The first is forward creep grazing where offspring are allowed into the next paddock in the grazing sequence before their dams. This occurs on many farms and ranches by default because the fences won't hold the young critters.

In this system the offspring benefit from the opportunity to selectively graze higher quality forage before their dams get there. This can work with grazing periods from one day to a couple of weeks. The key factor is the opportunity to select the choice forage first. True creep grazing creates a separate pasture of higher quality forage for the offspring to utilize but the parent animals never have access unless it is just to clean up the creep area.

Seeding small fenced areas with annual crops among the permanent pastures in the most common form of creep grazing. Perennial forages of higher nutritional value, such as legumes, can also be used, although the risk of bloat must always be considered with certain legumes.

For creep grazing to be beneficial to the calves or lambs, there must be a nutritional advantage in the creep grazing area and it must be readily accessible by the young animals. The nutritional advantage may be greater forage availability, higher nutritional quality, or more palatable forage. When pastures are grazed short and availability restricts forage intake, just allowing the offspring access to an area with a little greater forage availability can help their performance.

Livestock grazing permanent pastures such as endophyte-infected tall fescue or bahia grass, which are generally considered lower quality, are most likely to benefit from creep grazing because the greater the nutritional differential between the base pasture and the creep pasture, the greater performance response.

The first grazing study I was involved in when I came to the University of Missouri in 1981 was a creep grazing study. In that project we had high endophyte KY31 tall fescue, low endophyte KY31, orchardgrass-red clover, and smooth bromegrass-red clover pastures. Within each base pasture type, there were calves which either did or did not have access to creepgrazing. The creep pasture consisted of timothy and birdsfoot trefoil, both high quality forages.

If you read the previous paragraph, the results of the study should be fairly predictable. The greatest calf gain response to creep grazing occurred on the high-endophyte fescue while there was no creep response on the orchardgrass or smooth bromegrass pastures. Averaged over the four years of the study, the creep grazing advantage on the high endophyte pasture was about 40 pounds increase in weaning weight. Some studies have shown more, some have shown less. The cost for creating the creep pasture has to be weighed against the increased animal output.

One of the problems we encountered using the perennial creep pasture was keeping it in high quality condition. Young calves, particularly, may not supply enough grazing pressure to keep the pasture in good condition. That is why it is sometimes important to use the main herd to recondition the creep pasture for high quality regrowth. Another important component of creep grazing is ensuring accessibility for the offspring while excluding the dams. Creep access can be very easily managed with electric fences.

R.L. Dalrymple and the group at the Noble Foundation developed many interesting avenues for creep access. Using a series of pinlock insulators on either steel or wood posts is the easiest in my opinion. As calf size increases the electric fence height can easily be adjusted to accommodate growing calves. When it is time to restrict access, simply lower the wire back to the lowest insulator level.

Sliding spring clips on round fiberglass posts can also be used, but they require shutting the fence off for adjustment while the pinlocks do not. In Brazil I saw board creep access where 1x4s were slid in and out of a frame to change height access. Wooden creep gate designs were developed by Dr. Roy Blaser at Virginia PolyTech in the 1950s and '60s and are shown in many text books and guidesheets from that era. Calf or lamb access is really only limited by the imagination.

For the grazier who has every acre in excellent high quality pasture, creep grazing is not likely to show much benefit. For the majority of us who have infected fescue and other lower quality permanent pasture to deal with and are restricted on how much land can go into annual pastures, creep grazing offers a good opportunity to improve pre-weaning performance. As with everything else in grazing, the key to making it profitable is management.

The Basics:

- For creep grazing to be beneficial to the calves or lambs, there must be a nutritional advantage in the creep grazing area and it must be readily accessible by the young animals.
- Forward creep grazing allows offspring into the next paddock in the grazing sequence before their dams.
- Seeding small fenced areas with annual crops among the permanent pastures in the most common form of creep grazing.
- It is sometimes important to use the main herd to recondition the creep pasture for high quality regrowth.
- Creep access can be very easily managed with electric fences.

Think About Your Farm or Ranch:

- What would be the costs to implement forage creep grazing? Would this justify the increase in weaning weight?

Extending the Grazing Season

30 Managing Cow Condition Through Grazing

Understanding when to add weight and when to lose weight is a fundamental part of managing a year-around forage-based livestock system.

Fat cows, skinny cows, or in-between cows: do you always want one or the other, or is there a time when any of these can be acceptable?

Well, it depends.

How fat is fat and how skinny is skinny? And does time of year or stage of production make a difference?

We need to first be able to systematically describe skinny and fat. In the USA we use a 9-point scale to describe beef cow condition. Officially we call it the body condition score (BCS) system.

You will frequently see "BCS" used in publications with no further explanation of what it means. Other countries may use a different scale as do dairy cows in the USA.

On the 9-point scale, a "1" is skin and bones with no definable muscle structure. Usually you only see these ones as they're getting hoisted on the rendering truck.

A "9" is on the opposite end of the scale and is best described as obesity. These are the cows that the calves in the back pasture make "Your momma is so fat...." jokes about.

We don't see a lot of 9s in commercial cow herds. As a general rule, we usually work in the range of 4 to 7, with an occasional 3 or 8 being acceptable in the right circumstances.

Depending on cow frame score, one BCS unit will be equivalent to 70 to 100 lbs change in body weight. Basically it amounts to adding and removing body fat, which is nature's feed bin.

Most wild animals add body fat before winter and then lose a significant amount of weight over winter. It is an extremely economical system for supplementing energy during hard times. The same approach can be used in winter management of cattle, sheep, and goats. Understanding when to add weight and when to lose weight is a fundamental part of managing a year-around forage-based livestock system.

One of the nice things about the BCS system is there are well defined benchmarks to shoot for to achieve certain performance responses.

If cows are below BCS 5, the likelihood of them breeding to stay on a 365-day calving interval is greatly diminished. With each unit increase from 5 to 6 to 7, the anoestrous period shortens and a higher percentage of cows breed early in the breeding season. If the cow is gaining in condition, the effect is even better.

Even gaining from 4 to 5 can give better breeding results than sitting still at 5. As cow condition increases, winter feed efficiency increases due to the insulating effect of fat. So keeping cows in the proper condition at the right time of year is an important part of herd management. The question is how to do it efficiently and effectively.

As BCS at calving and breeding are the important ones to consider, planning when these events occur is pretty fundamental to success. If calving occurs at a time of year when it is easy and inexpensive to put on weight, it is pretty easy to ensure good breeding performance. Unfortunately the opposite is also true.

If calving occurs when it is difficult or expensive to put on condition, breeding performance is less likely to be acceptable, at a reasonable cost. The two easiest times to put weight on a cow are spring and fall, due to grass and weather conditions.

Winter is most difficult, except in the Deep South where cool-season annual forages are at their best during winter months.

In drought prone areas, summer can be a challenging time to put weight on. This leaves spring to early summer and fall as the best times to calve in most of the country, depending on winter conditions.

A big part of the inefficiency of most cow operations is they calve January to March, which are the three hardest months to maintain, let alone add, body condition.

A fat cow has economic value in the fall or going into a summer dry season in arid country. That economic value is only captured if you are willing to let her shed those pounds during the lean times.

A spring calving cow going into winter at BCS 7 can drop 150 or so pounds and still calve at BCS 5 when spring grass comes and do so on a pretty tight budget. Trying to keep her as a 7 all through winter and calving her as a 7 can be very expensive and does not lend itself to economic efficiency in a cow operation.

The key to making this system work is putting weight on the cow in the fall after weaning on cheap, high quality fall regrowth pasture or calving late enough in the spring or early summer that she can add weight back on cheap spring grass.

Stockpiled cool-season perennial pasture supplemented with cow fat makes for a very economical wintering program. Stockpiled warm-season perennials will likely need additional protein supplement to make the system work.

Understanding what it takes to put weight on and knowing when to lose it dictate grazing management decisions with the cow herd. If you follow the two basic rules of grazing, you can manage cow condition pretty well.

Rule #1: The more you eat, the fatter you get. If cows need to add weight, they need to increase their daily forage intake. How do you do that? Leave taller residuals after grazing, move more frequently, or allocate higher quality forage.

Rule # 2: If you want to lose weight, go on a diet.

If you are ready to let cows shed a few pounds, it is time to tighten up grazing. Graze to shorter residuals, if the pasture can stand it. Let the cows clean up the garbage after a higher performing group of animals has creamed the paddock.

These are the day-to-day grazing management decisions that affect cow condition. Knowing when to apply which tool is the key to managing cow condition.

The Basics:

- Keeping cows in the proper condition at the right time of year is an important part of herd management.
- The two easiest times to put weight on a cow are spring and fall, due to grass and weather conditions.
- As BCS at calving and breeding are the important ones to consider, planning when these events occur is pretty fundamental to success.
- The key to making this system work is putting weight on the cow in the fall after weaning on cheap, high quality fall regrowth pasture or calving late enough in the spring or early summer that she can add weight back on cheap spring grass.
- To increase body condition, leave taller residuals after grazing, move more frequently, or allocate higher quality forage.
- If you are ready to let cows shed a few pounds, it is time to tighten up grazing. Graze to shorter residuals, if the pasture can stand it.

Think About Your Farm or Ranch:

- Are your cows' calving in sync with their BCS?

31 Stockpile Forage for Low Cost Wintering

Making sure there is pasture available in the winter months is a job that begins in the spring and continues through the summer months.

From time to time, I am sure my neighbors think I'm a little wacko. Why on earth would a guy go out in a foot of snow and run polytape across a pasture?

Of course I wonder why on earth someone wants to start a tractor on a cold snowy morning just to haul expensive hay out to a bunch of lazy cows. Different strokes for different folks, I guess.

The basic difference is in attitude. Some people work for their cows, I prefer to have the cows work for me.

An important part of having the cows work for me is making them rustle their own grub during winter when pastures aren't growing in our part of the world. My job is to make sure they have something to rustle, not rustle it for them.

Making sure there is pasture available in the winter months is a job that begins in the spring and continues through the summer months. You can't have pasture for winter grazing just because you decide in October that you want to graze cows all winter.

Producing winter pasture is a combination of having the proper stocking rate at the right time of the year as well as several agronomic practices. Earlier, I talked about some of the practices we used on our farm to create a variable stocking rate across seasons to bring animal demand and forage supply into better balance. To quickly summarize: you need more animal demand when pasture is growing fast and less animal demand in the dormant season. Using a combination of contract grazing and timing of calving allows us to increase animal demand April to August and then reduces demand in late summer. The reduction in demand in August is what allows us to begin the process of stockpiling pasture for winter grazing.

In the Midwest and Upper South, stockpiling tall fescue-based pastures is the bread-and-butter of winter grazing. In other parts of the country and world, other forages make up the staple dormant season diet. Almost any forage can be stockpiled, but the results vary significantly in terms of forage availability, quality, and durability.

Several factors make tall fescue ideal for stockpiling. It is extremely productive in the fall. We have consistently achieved stockpile yields of 1.5 to 2 tons of forage, depending on late season precipitation. About one year in seven we will have no fall pasture because of dry conditions and about one year in four we will have less than we hoped for.

In response to cooler nights and shorter days, tall fescue accumulates high level of non-structural carbohydrates, which substantially increases the energy content of the grass so the quality of stockpiled fescue is surprisingly good. While we sometimes complain about the coarseness of fescue during the spring and summer, it is that very characteristic that makes stockpiled fescue very weather resistant and durable through the winter. Fescue also forms a very durable sod, keeping livestock out of the mud in our highly variable winter temperatures.

Growing the right kind of stockpiled pastures doesn't happen by accident but requires planning and management. Selecting the right pastures is a good starting point. Thin, weedy pastures do not stockpile well. If pastures are thin and weedy because of winter grazing, then changes in winter management need to be made.

While some winter treading damage is inevitable in a wet winter climate, the damage can be minimized through short grazing periods and not wintering on the same ground year after year. I try to avoid being on the same pastures in the winter for at least three years. This allows plenty of time for recovery of any damaged areas and keeps weeds to a minimum.

Some of our pastures just aren't suitable for winter use due to soil conditions, plant community, or landscape. These are pastures that can be rested and used during the late summer while winter stockpile is accumulating on other pastures.

We like to start our stockpile with all new regrowth so pastures are either grazed or clipped to a three to four inch residual to begin. In our environment, we will achieve the maximum potential stockpile yield in about 75 days of growth, given typical late summer-early fall conditions. We generally figure active fescue growth ends around November 1. Backing up 75 days from this date suggests that about August 15 is the optimal date for us to begin stockpiling.

The 75-day rule applies to a lot of different cool-season forages in many environments and is a good working rule of thumb. Letting pastures accumulate longer than this will not produce any higher yields but will significantly lower forage quality.

When wintering dry, pregnant females, maximum stockpile yield is usually the goal. If wintering cows or ewes giving birth in the fall or growing out young stock, higher quality forage may be desired. Shortening the stockpiling period will produce higher quality pasture, but gives lower yield.

Nitrogen is essential for growing fall pasture. Legumes, manure, or fertilizer can all be used to supply N. Tall fescue-legume mixtures can yield as much as pastures receiving about 60 lb N/acre, but pasture quality will deteriorate much faster than N-fertilized grass.

Planning the winter use period to utilize grass-legume

stockpile first and save N-fertilized grass for late winter helps offset this challenge.

Red clover is the best legume to use in stockpile systems as it has moderate N-fixation capacity, good fall vigor, weathers reasonably well, and has the capacity to reseed itself following the stockpile grazing. For many years the standard N recommendation for stockpiling has been to apply 40 to 60 lb N/acre in mid-August. With ample fall rain and intensive strip grazing, N rates up to 100 lb/acre can be economically feasible in the Midwest and Upper South.

We have stockpiled many other types of pasture including orchardgrass and smooth bromegrass, as well as very diverse mixtures. Used fairly early in winter, these other pastures do okay, but they don't have the staying power of tall fescue pastures to still have good yield and quality late in winter.

What we have found over the last several years is a pasture does not have to be 100% fescue to stockpile well. With 30 to 40% tall fescue in the mixture, we see the fescue providing protection to a lot of the other forages and making diverse pastures work reasonably well for stockpiling. The increased diversity offsets fescue toxicity through the spring and summer months. Stockpiling native range is the wintering base for many low-cost cow-calf ranches in the arid West, as well as countries like Australia, South Africa, and much of South America. Using winter rangelands will be discussed later.

The Basics:

- Almost any forage can be stockpiled, but the results vary significantly in terms of forage availability, quality, and durability.
- Growing the right kind of stockpiled pastures doesn't happen by accident but requires planning and management.
- The 75-day rule applies to a lot of different cool-season

forages in many environments and is a good working rule of thumb.

- When wintering dry, pregnant females, maximum stockpile yield is usually the goal.
- Shortening the stockpiling period will produce higher quality pasture, but gives lower yield.

Think About Your Farm or Ranch:

- Using the 75-day rule, what is the optimal date in your area to begin stockpiling?
- Which forages can you utilize for stockpiling?

32 Use Annual Forages to Extend Winter Grazing

Cool-season annual forages are another option for extending the winter grazing season, sometimes on both ends of winter!

Perennial pastures usually offer the lowest cost option for winter grazing dry, pregnant live stock, but may not be of adequate nutritive value for growing animals or lactating stock during the winter months.

Cool-season annual forages are another option for extending the winter grazing season, sometimes on both ends of winter!

Cool-season annuals include small grains like wheat, rye, triticale, barley, and oats; annual ryegrass; legumes like crimson and arrowleaf clover, winter peas, and vetch; and brassicas like turnips, kale, rape, and swedes. The common feature of cool-season annuals is that they must either be seeded, or allowed to naturally reseed, for emergence in late summer or fall.

Some will survive the entire winter and be productive in the spring while others may only provide forage in the fall and early winter.

In most cases winter annuals are seeded in a prepared seedbed and require substantial fertilizer inputs which can make them appear to be fairly expensive alternatives.

For example, establishing annual ryegrass with appropriate fertilization and seeding rate may cost $50 to $80 per acre compared to spending $20/acre to apply N to tall fescue for stockpiling.

While the per acre cost is much higher for the ryegrass pasture, total forage production may be greater for the annual pasture and individual animal performance will be higher.

But the real comparison to make is what is the cost for the next-best alternative forage and will it provide the target level of performance?

In a recent study at FSRC comparing alternative forage systems for fall calving cows, we found the per day feed cost per cow-calf pair for stockpiled pasture and a mixture of annual ryegrass and cereal rye to be 31 cents and 61 cents per day, respectively, compared to $1.32/day for cows on hay.

Yes, the annual pasture cost nearly twice as much as the stockpiled tall fescue, but the annual pasture was still less than half the cost of feeding hay.

On farms and ranches where row crops are also produced, using annual pastures is very easy.

On rolling land that is entirely permanent pasture, using annual crops is much more challenging and may not be a viable alternative.

Which winter annual or combination to use depends on location, type of livestock, and window of use. Investing in an annual crop that provides only front-half forage usually makes the per head per day cost higher compared to a forage that provides grazing on both the front and back half of winter.

At our location in north Missouri, we are on the frontier of reliable winter survivability of annual ryegrass. In a severe winter when annual ryegrass does not survive, our per day cost may double because of the lost spring production.

When winter survival is not reliable, using a highly productive brassica, such as kale, may be more cost effective.

Knowing the likelihood of winter survival is one of the most important aspects of using winter annual pastures. Because of their high quality potential, annual pastures may be better suited for growing or lactating livestock than for maintenance of mature animals.

Annual pastures can extend the high quality grazing season for grass-based dairies by 60 to 90 days in many parts of the country. Annual pastures also offer opportunity for producing pasture-finished beef and lamb more months of the year in what is currently a very seasonal market.

Some winter annuals produce substantial forage on both sides of winter. Annual ryegrass, cereal rye, and triticale are good examples.

Most of the brassicas produce forage only on the front half of winter while winter annual legumes tend to produce very little on the front half but high yields in spring.

Planting a mixture of two types of annuals can provide forage from the same acre on both sides of winter.

Because of the monoculture mentality commonly associated with winter annual pastures, the value of planting mixtures is often overlooked. We have used mixtures of oats and rye and rye and turnips to provide greater forage production in the fall.

The oats are expected to winterkill, but they are the highest yielding pasture of the small grains when sown in late summer.

The key to using mixtures is utilizing the two crops in different windows of opportunity.

Because of the relatively high cost of annual pasture, it is critical to manage grazing intensively to maximize their cost effectiveness.

We have found that a 3-day strip graze will provide about 70% utilization rate while a 14-day allocation may provide only 50% utilization and much lower than that in poor weather conditions.

To illustrate the impact of utilization rate on daily feed cost let's consider a pasture producing 8000 lb/acre at a cost of $60/acre. At 50% utilization, we are harvesting 4000 lb of forage at a cost of 1.5 cents/lb. If it takes 30 pounds of this forage to feed a fall-calving cow each day, the per day cost is 45 cents/cow.

If utilization is pushed up to 70% through tighter strip grazing, we will harvest 5600 lb forage/acre and the cost per pound drops to 1.07 cents and the daily cost is 32 cents/cow.

To put it in another perspective, we generally figure it takes about 10 minutes to move a poly-fence for strip grazing. We can figure the four additional moves in each two week period for the more intensive strip grazing will take about 40 minutes.

With 100 cows in a herd, saving 13 cents/cow/day adds $13/day to the net return for the herd or $182 for each two week period. If it only took 40 minutes to realize that additional gain, the hourly wage for moving fence is $272.

Besides their cost and risk of winterkill, two other challenges to using annual pastures are that it may either be too dry or too wet. Because the normal seeding time for cool-season annuals is August through September, it is possible to face establishment failure due to drought. While seed of some species may lay dormant over winter and emerge with spring rains, others may sprout with just a little soil moisture and then dry out when no additional rains come.

Based on long term weather records, it is fairly easy to calculate a stand failure risk factor for your area. For example, in north Missouri we expect dry enough conditions for stand failure during our seeding window about one year in seven.

The other challenge is excessively wet weather that leaves fields too soggy for grazing without creating one helluva a mess. This is actually a higher risk factor for us than are the dry conditions.

Over the last several winters when conditions have been much milder than the norm, muddy conditions have limited our opportunity to graze pastures in a timely manner about one year in three. We do end up grazing the pastures, but not when they are at optimum yield and quality.

Spring tillage to clean up the soil damage is sometimes

required.

While winter annual pastures aren't for every operation, they do offer another forage alternative for producers needing higher quality forage than stockpiled perennials can provide. Compared to feeding harvested forages, winter annuals usually pencil out as a cost effective alternative.

Remember that intensive utilization is the key to making them most profitable.

The Basics:

- Which winter annual or combination to use depends on location, type of livestock, and window of use.
- When winter survival is not reliable, using a highly productive brassica, such as kale, may be more cost effective.
- The key to using mixtures is utilizing the two crops in different windows of opportunity.

Think About Your Farm or Ranch:

- Compare the cost for your next-best alternative forage. Will it provide the target level of performance you need?
- Using long term weather records, calculate a stand failure risk factor for your area.

33 Snow Grazing for You Grass Farming Crackpots

It's all a matter of perspective.

If you want to be a cutting edge grazier, one of the first things you have to learn to accept and live with is that your neighbors will probably think you're a crackpot. But you know, it's all a matter of perspective.

I am sure some neighbors think I am absolutely crazy when they see me unreeling polytape across a field covered with a foot of snow. From my perspective, I think anyone who runs a $40,000 tractor several hours every day to feed a bunch of cows they are losing money on must be a crackpot. So it's all a matter of perspective.

To me it is much easier to use some polytape and step-in posts that cost me less than $100 and spend 15 minutes feeding a bunch of cows. We have repeatedly shown at FSRC that cows can be wintered on pasture for about 1/4th the cost of feeding hay. Most serious graziers believe what I am saying and so do a lot of "conventional" farmers but they still have their doubts that it will work for them.

By the way, have you ever thought about how unconventional "conventional" farming really is? I want to revisit some of the questions I get asked on a regular basis about grazing in the winter time.

"How do you make cows graze through snow when they're used to eating hay all winter?"

The first and obvious answer is don't feed them any hay. Hungry cows learn to root or die pretty quickly. It is nice to be concerned about the poor cow who doesn't know how to graze but it doesn't take even the most finicky barn-raised Holstein very long to figure out what is going on.

Our job is to make sure that there is accessible forage under the snow and ice. Four inches of bluegrass under 12 inches of snow is a challenge to the cow. A sheep can live all winter on the same kind of pasture that a cow would starve on. Doing a proper job of developing stockpile, winter annuals, or swath grazing to ensure that they have forage available is where our management is critical. Different classes of livestock have different requirements and we need to plan the winter pasture program to meet those needs. If you fail to do this, then winter grazing won't work.

"How do you get those step-in posts into the frozen ground?"

There are some step-in posts that simply will not work in the wintertime because you can't get them in the ground. The right type of post has a slender spike that meets less resistance for penetration, a broad step so you can get your icy boot to stay on the step, and a lot of flexibility even at sub-freezing temperatures. Most step-in posts have too thick a spike, too small a step, and become too brittle in winter to be usable. The technique for putting a step-in post into frozen ground is similar to putting it in dry hard ground.

Don't try to just shove down and expect the post to go. A gentle rocking motion with the foot accompanied by downward pressure usually gets the job done.

Another thing to remember is that you don't expect the fence to be up for a month, just for a few days. If you only get a couple of inches of penetration, that is good enough. My experience has been we can work pretty effectively with a good step-in post down to about -10 degrees F. Below that temperature, the soil has expanded to the point that there is no give left in it. At this point you can opt for allocating bigger areas and moving fences less frequently. If you do this, feed wastage will rapidly increase and you will be back to feeding hay sooner.

We have used either chunks of wood with holes drilled in them to accommodate the post spike or concrete rounds with a tube in the center. These will support step-in posts for a few days until you move to the next strip. Specifically we use 12 inch square X 3 inch thick bridge plank with a hole drilled that holds the post snugly in place. The step-ins we use have 3/16 spikes so we use a 7/32 bit.

The concrete rounds are made in little forms by slicing 5-gallon buckets into 3-inch rings. We set a piece of 3/16 tubing in the center to accommodate the spike.

"My tape gets coated with ice and it goes to the ground."

Yep, that sure happens sometimes. If cattle are well broke to electric fence, they probably still won't cross it. If they get pushed by the storm then they are going across anyway and it probably wouldn't matter if the tape had been up or not. When tape does get ice encrusted, it is quite simple to remove the ice.

This is a trick I learned just last winter but works real well. With a spare step-in post, just hook one of the clips around the tape and start walking with firm pressure held on the tape. The ice will pop right off. We were running several 400 foot strips last winter and had ice on the tapes several times. I found that I could clear a tape in less than two minutes. For those of you who know me, you know that I have a bad habit of timing exactly how long it takes to do specific tasks. Use one of the lower clips that is rarely if ever used for fence to do this job just in case you do break it off.

"Do I need to use a back fence?"

As a general rule there is no need to use a back fence when grazing stockpile or swaths. The pasture is dormant and you can't really hurt it. Winter annual pastures may need a little more attention regarding back fencing as mid-afternoon warmups can make pastures really squishy and winter annuals are highly susceptible to trampling damage. We generally start grazing stockpile closest to the winter water source and just work our strips away from water.

"Can cows live on snow as their only water source?"

This is fairly controversial and depends on stage of production and dryness or wetness of the snow. I have had many people argue that snow is snow and it contains the same amount of water. Yes, a pound of snow is a pound of water. The real issue is how much do they get in their mouths? Cows do not get high water intake on pasture by licking up snow. They get it from ingesting snow as they graze forage. A wet, heavy snow (good for snowmen and snowball fights) tends to stick to the forage and gets consumed at a fairly high rate.

Dry powder snow (good for skiing) blows and drops away from the forage and water intake is low. I have run dry cows grazing stockpile for extended periods without supplemental water and gotten along fine with it. As soon as they were shifted to hay, they needed water because they were not licking snow. Fall and winter calving cows that are lactating have higher water requirements and probably can't meet their needs just from snow.

Winter grazing does have a few more challenges than summer grazing but coping with it is largely a matter of having the right attitude. What is conventional or crackpot is just a matter of perspective.

The Basics:

- To get cows to learn to graze through snow, don't feed them any hay.
- Remember, your fence only has to be up for a little while.
- Doing a proper job of developing stockpile, winter annuals, or swath grazing to ensure that they have forage available is where our management is critical.

- Different classes of livestock have different requirements and we need to plan the winter pasture program to meet those needs.

Think About Your Farm or Ranch:

- How can you plan your winter pasture program to meet the needs of your classes of livestock?

34 Wintering on Native Range and Supplementation

The capability to graze year-around is driven much more by attitude than environment.

Long before the white man came to the Prairies and Plains with his plows and European grasses, hundreds of millions of acres of North America was covered by vast expanses of grasslands and savanna. Bison, elk, deer, and pronghorns numbering in the millions roamed the grasslands and lived out their lives here.

When the rains came regularly, they grew fat in the summer and fall and grew lean over the winter, but were rejuvenated when spring grass came. When the rains didn't come, they grew lean in the summer and many of them died in the winter.

Then white man came and replaced the wild herds with cattle and sheep. They first brought horses and horse-drawn implements to make hay and grow grains. Then came the tractor and diesel fuel and more and more of the native grasslands disappeared to be replaced by tame grasses and row crops.

Technology allowed us to insulate our domestic livestock from the uncertainties of life the native ruminants experienced. Our cows did not need to die in the winter for lack of forage for we could provide it for them. But it all came at a cost, and everyday the cost of protecting our livestock from hunger grows higher.

Over the last ten years both producer experience and university research have shown it to be much less expensive to winter cattle on stockpiled pasture than to feed them hay. Grazing beef cows during the winter in the temperate region has about one-fourth the cost of feeding hay. In the fescue belt, the big advantage we have is a cool-season grass that grows well in the fall, maintains adequate protein level, and weathers well.

Wintering on native range offers greater challenges than wintering on stockpiled fescue. Through a large part of the grass growing regions of the USA, warm-season grasses make up the bulk of the rangeland. Unlike tall fescue, native warm-season grasses do not grow well in the fall, do not contain adequate protein levels, and may not weather very well. In quick summary, it is not nearly as easy to winter cattle on rangeland as it is on fescue.

Should we just head back to the hay fields and spend some more money?

I don't think so. Winter grazing on native range just takes a little more management and a few more inputs than wintering on fescue, but it is still much more cost effective than feeding hay.

As with any dormant-season grazing program, the first requirement is to make sure you have some rangeland available to accumulate forage during the growing season to be grazed later in the winter. This means rest is required for part of the ranch for part of the summer.

Rangeland in this country exists in a wide range of environments from the extremely wet and humid Gulf Coast region to high desert rangeland in Nevada receiving less than six inches precipitation annually. Obviously the management strategy for providing winter forage is going to vary greatly.

A 30- to 40-day rest period may be adequate in Louisiana while a full year's deferment may be needed in Nevada. The really cool thing about these extremes is that I personally know ranchers in both of these locations who graze year-around while their neighbors feed hay for several months. Everywhere I travel, I find the capability to graze year-around is driven much more by attitude than environment.

Flexibility in year-around stocking rate management is one of the keys to being able to provide the necessary rest periods to accumulate winter range. In public-lands ranching operations, gaining the flexibility in timing and numbers to optimally utilize winter grazing is much more challenging than grazing entirely on private land.

While changing attitudes in the ranching community seems challenging, changing bureaucratic attitudes is monumental. But not impossible.

In drier areas, using a strategy of light grazing early in the season followed by resting the range the remaining half to two-thirds of the growing season has been shown to provide better quality winter range than deferment for the entire season. However, if a range site is grazed for over half of the growing season, winter forage yield may be quite low. If winter forage supply is marginal and the cattle are allowed to over utilize it in the winter season, spring growth will be severely diminished.

Because of the low protein content of winter-dormant warm-season grasses, some supplementation is almost always necessary even for dry stock. Fall-calving on dormant range takes top notch management and a nutritionally balanced supplementation program. The one great advantage fall calving offers is having a weaned calf to hit spring grass in prime condition.

Most ranchers who do a good job of winter range utilization calve in late spring and early summer. This timing really minimizes nutritional requirements over winter and allows cows to regain some body condition on spring grass before they calve. Drs. Don Adams and Dick Clark at the University of Nebraska have clearly shown the operational feasibility and increased profitability of grazing winter range and calving late spring-early summer compared to "traditional" hay feeding programs with February-March calving.

Supplementation with a high protein, fiber-based byproduct feed greatly enhances the suitability of low quality native winter range for beef cows. Corn gluten feed and distillers dry grain are excellent choices in more northern regions. Cottonseed cake is an excellent supplement in southern range country.

Remember the rumen bugs need to be fed protein to be able to ferment the fiber in dormant grasses. A byproduct-feed supplement with 20 to 25% crude protein content can be fed to dry cows at one-quarter to one-half percent of bodyweight per day.

One of the nice things about protein supplements is they don't need to be fed every day. Feeding a larger dose of protein every two or three days provides the same benefits as feeding a smaller amount every day.

Supplementing with high starch-low protein feeds like corn or milo don't do nearly as much good as fiber-based byproduct supplements. Alfalfa hay makes a good supplement for winter range diets also. Feeding six to eight pounds of good quality alfalfa hay per cow provides appropriate supplemental protein.

Monitor cow body condition score regularly through fall and winter to make sure cows on winter range aren't losing too much condition too fast. If you always know which direction your cows are heading, you can step up the supplementation rate or change the protein level to avert a wreck. It is always much cheaper to prevent a problem than it is to fix the wreck.

Wintering growing stock on winter range requires higher levels of supplementation than dry cows. The younger animal's rumen is not as well developed for utilizing coarse forage and their basic metabolic requirements are higher so the difference between those requirements and what dormant grass can provide is greater.

Young stock may be supplemented from about one-half

to one percent of their bodyweight per day. As with the mature cows, alternate day feeding works satisfactorily.

If you are located in an area with extreme cold, energy requirements will be higher as the temperature falls below freezing. Additional energy may need to be supplemented along with protein. Planning your winter grazing for more sheltered areas such as sheltered valleys or south facing slopes can help reduce winter's impact on nutritional requirements.

Because of the large expanses of land used in winter range conditions, tight strip grazing like we practice in the fescue belt is not as feasible. Regulating range allocation on a one to two week basis will make winter range last longer and keep quality higher in the undisturbed areas.

There are large-scale western ranches that use portable electric fencing to control winter grazing. Larger wildlife species like elk and moose play a little more havoc with fence than do our little whitetails. That is whitetail deer, not cottontail rabbits.

Getting wildlife used to your fences year-around is the first step towards gaining their respect. An elk does not like to get shocked anymore than a cow does. Maintain high power in the fence at all times and use durable materials to get the attention of big game species. I have seen polytape used effectively for stockpile grazing at Pinto Ranch, which borders Grand Teton National Park and experiences a great deal of elk overflow from the park.

Winter range grazing may take some changes in cow management and land use patterns to become a reality for many ranchers, but it offers good opportunity to significantly lower overwintering costs in rangeland environments.

The Basics:

- It is much less expensive to winter cattle on stockpiled pasture than to feed them hay.

- Winter grazing on native range just takes a little more management and a few more inputs.
- Make sure you have some rangeland available to accumulate forage during the growing season to be grazed later in the winter.
- Flexibility in year-around stocking rate management is one of the keys to being able to provide the necessary rest periods to accumulate winter range.
- Most ranchers who do a good job of winter range utilization calve in late spring and early summer. This timing really minimizes nutritional requirements over winter and allows cows to regain some body condition on spring grass before they calve.
- Wintering growing stock on winter range requires higher levels of supplementation than dry cows.
- Planning your winter grazing for more sheltered areas such as sheltered valleys or south facing slopes can help reduce winter's impact on nutritional requirements.

Think About Your Farm or Ranch:

- Can your landscape help reduce your animals' winter nutritional needs?
- Do you need to change your calving season?
- What is a cheap source of protein supplement in your area?
- Plan your grazing strategy to allow adequate rest periods for the pasture and a stockpile of forage for the winter.

Animal Care

35 Pasture Weaning for Health and Weight Gain

Pasture weaning allows the calf to enjoy a diet it is used to in a familiar environment with only the absence of mom to adjust to.

For a lot of cow-calf producers, weaning consists of pulling calves off the cows the morning of sale day, shoving them on the trailer, and hauling them to town. With each passing season, the value lost on each set of calves escalates.

When I first got into the cattle business, the watch words were, "There are no premiums in the beef business. You just have to avoid the discounts." Those days are just about gone now.

There are premiums in the beef business and the discounts get heavier if your calves don't fit what the market demands.

What the market is demanding today is source-verified, healthy calves that have been effectively weaned and are ready to move on into backgrounding and finishing systems.

What if you are going to keep your calves and finish them on grass? Don't you also want calves that stay healthy and move right on into the next phase of your program?

The reason so many cattlemen avoid a true weaning process is because they're scared.

Scared that calves or cows are going to get out on the

highway; scared that calves will get sick; scared that calves will lose their bloom and won't sell as well as the day they came off the cow.

The old mentality is that you have to have a fortress to wean calves. Six-foot-tall board fences with barb-wire entanglements on top and land mines to keep the cows from breaking down the gates.

As with almost everything, it is only as hard as you make it.

Why is weaning so hard for some operations? Because of stress placed on the calf, cow, and operator.

Stress comes to the calf from four directions in "conventional" weaning strategies.

First we yank the calf away from the cow, usually through a clanging and banging corral and headgate. During the process we may castrate, dehorn, and inject the poor animal.

The calf ends up in a confined lot looking at something in a feedbunk that he doesn't even recognize as being food. This is pretty shocking for someone who has only known an open pasture, grass to eat, and always had mom to keep him out of trouble.

We have just inflicted social, physical, nutritional, and environmental stress on the animal. No wonder so many calves get sick at weaning.

I once heard a commercial weaning lot in our area described as a place to send healthy calves to make sure they get sick. There is a healthier, easier way to wean calves.

Pasture weaning allows the calf to enjoy a diet it is used to in a familiar environment with only the absence of mom to adjust to. With across-fence weaning, the calf can still enjoy mom's presence.

There are three main ingredients to successful pasture weaning: having cattle well broke to electric fence, good quality pasture at weaning time, and forward planning.

On most MiG farms or ranches, cattle get accustomed to electric fence early in life and learn to respect it. The only real failures I have seen with pasture weaning have been when cattle were not taught to respect fence early in life and fence maintenance was sloppy.

For spring-born calves in the tall fescue belt, stockpiled tall fescue makes excellent weaning pasture. In other areas annual ryegrass or small grains pasture are other alternatives.

For fall-born calves weaned in spring almost any pasture except infected tall fescue work well for weaning.

The key to keeping ADG high during the weaning period is healthy calves allowed to selectively graze high quality forage. On stockpiled pasture, we only expect the calves to remove about 1/3 of the forage. Grazing deeper will depress performance.

On ryegrass or small grains pasture, utilization can be pushed to about 50% and still maintain good results.

Forward planning is required to make sure you have the right pasture conditions in the right place at the right time. Also plan ahead to reduce physical stress of weaning by using polled bulls and castrating at birth.

We began pasture weaning at the Forage Systems Research Center in 1985. Since that time, over 3800 calves have been through the program with a total number of two sick calves. Not 2% and not two per year, but two sick calves in 18 years out of 3800-plus calves.

That is staying pretty healthy.

During a typical 21-day weaning period our calves usually gain around 1.5 lb/day without any supplemental feed or medication.

All calves are vaccinated pre-weaning with appropriate boosters given on weaning day.

Castration is done at birth and if any dehorning is needed, it is done with electric dehorners at spring roundup or several weeks after weaning.

On the Gerrish farm we don't even run cattle through the chute on weaning day but just gently sort them as they come up a paddock alleyway, steering cows to one side and calves to the other.

We do across-fence weaning, leaving cows and calves across a multi-wire electric fence for about three days and then moving them away from each other.

By the way, we have never had a sick calf at weaning either.

Based on our local sales, a calf that has been actually weaned and certified vaccinated is worth from $3 to $10 per cwt more compared to the calf that was hauled in bawling that morning. That is anywhere from $15 to $60 per head more even if the calf gained no weight during the weaning period.

An additional gain of 30 pounds made during a 21-day weaning period has been worth about $30 per calf here lately, more than enough to pay for any additional costs incurred. That is a lot of money to leave on the table.

The Basics:

- There are three main ingredients to successful pasture weaning: having cattle well broke to electric fence, good quality pasture at weaning time, and forward planning.
- The key to keeping ADG high during the weaning period is healthy calves allowed to selectively graze high quality forage.

Think About Your Farm or Ranch

- Do you have the right pasture conditions in the right place at the right time for pasture weaning?
- Have you planned ahead to reduce physical stress of weaning by using polled bulls and castrating at birth?

36 Wintering on Pasture: Rough 'em or Push 'em?

All the issues of forage availability and nutritive quality and their relationship to intake need to be taken into management consideration.

There are two basic approaches to wintering spring-born calves after fall weaning: push them hard or rough them through. Which to choose depends on where and when you plan to market them, their weight at weaning, expected price differentials, feed cost, weather risk factors, and a host of other factors.

The "push them hard" philosophy is based on getting cattle to slaughter weight at a young age and is typically done in confinement feeding situations. Depending on location, winter annual pastures offer another avenue for "push them hard" personalities.

Intensively managed annual ryegrass can generate rates of gain comparable to any backgrounding lot and many feedlots. Both research reports and producer experience confirm a 2.5 to 3.5 lb ADG. Annual ryegrass feed costs per pound of gain may be less than 20¢ compared to many confinement backgrounding programs at 35 to 50¢/lb.

The key to pasture-based backgrounding program is maximizing intake of high quality forage. All the issues of forage availability and nutritive quality and their relationship to intake need to be taken into management consideration. Other cereal grains such as wheat, rye, barley, oats, or triticale can also be used in wintering programs.

Establishing dense swards and providing adequate fertility to keep the grass coming are essential to making the most out of winter annual pastures. Muddy winters can limit the usefulness of winter annuals in some locations and seasons. Because calves are lighter weight than cows, the risk of mudding up the pasture is a little less with the younger cattle.

The "rough them through" approach assumes that if you own cattle long enough and put a little weight on over the winter, compensatory gain will kick in when spring comes and overall cost of gain will be acceptably low to allow a profit.

On borrowed money, the extended period of ownership may make this approach a little too expensive for some folks. If operating on equity capital, the rate of return still beats the stock market. This approach is probably more common in the fescue belt than anywhere else in the country.

Typical stockpiled fescue in the Midwest can provide 0.5 to 0.8 ADG if grazed to appropriate residual. As long as calves are not forced to consume the entire stockpile and are left with the opportunity to selectively graze, they can maintain a low rate of gain.

With well grown stockpile that is free of dead summer annual grasses and has a minimal amount of dead fescue leaves, ADG can exceed a pound/day in the northern fescue belt. In the upper and mid-South regions, rate of gain can be substantially higher using a leader-follower system with dry cows cleaning up the residual after calf grazing optimizes forage use. Adding supplemental feed can push up rate of gain to substantially higher levels.

As long as the value of additional gain exceeds supplemental feed cost, this can be a practical alternative to real basic "rough them through" management. Choosing the right type of supplement and feeding the appropriate level are key factors.

The fiber-based by-product feeds like corn gluten,

distillers dry grain, or soy hulls make better pasture supplements than just corn or other grains. Starch-based feeds alter the rumen environment and lower the digestive efficiency of forage fiber. Starch-based supplements fed at rates higher than 0.3% of bodyweight can actually depress total dry matter intake and result in disappointing results.

The fiber-based concentrate feeds do not affect rumen function and result in better use of forage fiber. By-product feeds are frequently fed at rates up to 1% of bodyweight in stockpiled tall fescue backgrounding systems.

I want to come back to the economics of supplementation and give a word of warning to folks dealing with some feed dealers for the first time. Here is a presentation technique that I have seen used way too many times to sell a cattle grower on a supplementation program.

Feed dealer, "Feed our supplement and your calves will gain 2 lb/day instead of the 1.25 lb they are doing now."

Cattle Bob, "What's it gonna cost me?"

Feed dealer, "Well it's $180/ton. That's just 9¢/pound. Feed six pounds/day. That's just 54¢/day. You get 2 lb/day gain so it just costs you 27¢/pound of gain. That's pretty cheap gain."

So what's wrong with this picture?

The calves were already gaining 1.25 lb/day without the supplement, so all you are really getting is the extra 0.75/day. Feed costs at 54¢/day makes the additional gain cost 72¢/lb, not the 27¢ the feed salesman claimed. The additional feed can only be spread over the marginal gain.

This little song and dance routine is why a lot of cow-calf producers don't believe it pays to wean their calves. They were sold a high-priced weaning feed that didn't provide enough additional gain to pay the feed bill. Constructing a by-product feed supplement and feeding at the appropriate rate can give a very profitable response, but be careful how you analyze the economics of the supplement.

Any of the approaches to wintering weaned calves

described above can work. High cost pasture or supplementation programs require high rates of gain to make them profitable. Maximizing forage intake while keeping supplementation costs under control are keys to making the "push them hard" approach work. Putting on lots of pounds over an extended period of time with really low cost pasture options is what makes "rough them through" work.

The Basics:

- There are two basic approaches to wintering spring-born calves after fall weaning: push them hard or rough them through.
- The key to this pasture-based backgrounding program is maximizing intake of high quality forage.
- Establishing dense swards and keeping adequate fertility to keep the grass coming are essential to making the most out of winter annual pastures.
- Choosing the right type of supplement and feeding the appropriate level are key factors.
- Be careful how you analyze the economics of the supplement.
- Maximizing forage intake while keeping supplementation costs under control are keys to making the "push them hard" approach work.

Think About Your Farm or Ranch:

- Push the pencil to see which works best for your resources, environment, and management style.

37 Dealing with Pests and Parasites on Pasture

All of us have some issues to deal with.

Whenever you pick up any of the popular press magazines dealing with livestock you can count on seeing a number of glossy advertisements warning about the millions of dollars we are all losing to pests and parasites that afflict our livestock.

Flies, ticks, worms, lice and all their other *compadres* are out there sucking, biting, and burrowing on all of our critters. Or are they?

Few things are as widespread and still as site specific as pests and parasites. All of us have some pest issues to deal with but none of us have all the problems.

It is real easy to spend too much money on pest management, but it also easy to spend too little in the face of a real problem. The first step in dealing with pests and parasites is to make a realistic assessment of what are likely to be problems on your place with your livestock.

Flies occur everywhere in North America and, for all practical purposes, everywhere in the world. The question to answer is are they an economic problem on your herd in your environment?

One recent advertisement says that horn flies cost the beef industry over $800 million annually. Dr. Rob Hall conducted horn fly control research at the University of Missouri for over 30 years and his conclusion was that north of the Missouri River it didn't pay to try to control horn flies. South of the river was a different story.

Why the difference?

Look at the magazine advertisements and the flyers that come in your mail and you can be convinced that every cow in the country should be dewormed because they will wean heavier calves if parasites are controlled. The late Dr. Bob Corwin conducted internal parasite research at the University of Missouri for over 20 years and arrived at a similar conclusion regarding stomach worms. There was no point in deworming mature cows north of the Missouri River.

Once cows got past a certain age, deworming had little impact on the weaning weight of their calves. First calf heifers are a different story though.

Both heifers and calves benefit significantly from deworming. The bottom line is the same as many other issues discussed here. Your situation is almost always unique and you need to deal with your issues, not generalities.

Environmental factors affect what pest and parasite problems you are likely to encounter and how severe a problem they will be. Why was there a difference in the economics of fly control between north and south Missouri? Length of the pest season is one factor.

The longer your warm, comfortable weather lasts, the more problems you are likely to have. More generations of pests can cycle in a year. Survivability can be greater.

Horn flies, for example, hatch in about 30 days when the temperature is around 60 F° but every 10 days when the temperature is over 80 F°.

The greater preponderance of endophyte-infected tall fescue in south Missouri makes cattle more susceptible to secondary stresses like parasites. These are the types of environmental factors you need to consider as you develop your pest management strategy.

Variations in temperature, humidity, toxic plants, and duration of winter will make your strategy different from what you see in the magazines. Grazing management also affects pest and parasite management.

There is a popular myth that rotational grazing breaks parasite cycles and eliminates the need for parasite management. Long term rotations that change livestock species use can break parasite cycles, but just rotating cattle every 30 days through ten paddocks on a hundred acres does very little to reduce a parasite or pest problems.

Because the parasite life cycles vary with temperature and moisture, they are not as predictable as your grazing cycles. It is just as likely that you will be moving stock back into the pasture at peak hatch as it is unlikely.

We conducted several grazing and parasite studies in cooperation with Dr. Corwin. What we found to be important management factors to reduce parasite load on pastures was leaving a taller residual and reducing stocking rate.

In side-by-side comparisons of rotational and continuous grazing at equal stocking rates, we found virtually no difference in parasite loads with the different grazing management systems. However, regardless of grazing system, we always found higher parasite loads with higher stocking rates. We believe this is due to lower residual heights at high stocking rates which led to much higher re-infection rates as cattle grazed closer to the ground.

Most graziers know that worm larvae hatch out of dung plies and move up onto forage plants with dew droplets. What most don't know is that very few of those parasites ever get more than just a few inches up into the canopy.

Grazing to four-inch residual reduces the likelihood of reinfection to almost zero compared to grazing to two inches. Rotational grazing makes very little difference, residual management makes almost all the difference.

One of the other observations that Dr. Corwin made over his many years of research was that within every grazing treatment there were always individuals who never showed any infection and there were others who would become highly infected. The conclusion, and it has been echoed by researchers and breeders around the world, is there is a very strong case for genetic resistance to parasites.

Here is a common ranch scenario. We have a hundred cows and we have just weaned the calves. Half a dozen of the cows just look pitiful and we decide it is due to stomach worms.

Now we have a couple of choices we could make.

One is deworm all of the cows because in the advertisements it says make sure you treat all animals to eliminate any carriers in the herd. This is what most managers have historically done.

However, if records were kept it would soon become apparent that it is the same cows each year that need the deworming. We have been spending $500 to $1000 to deworm 100 cows because six needed it. The added production from those six won't pay the bill for the other 94 head.

The better choice is to cull those animals that get plagued by parasites year after year. Over time the genetic base of your herd becomes more and more parasite resistant and the need for external inputs for parasite control disappears. Bottom line here is cull the problem animals and work toward a better adapted genetic base.

Similar to the worms, I have observed over the years in my own cows that there are individuals who never have more than 20 or 30 horn flies on them while a mate right next to them has several hundred. Though I never took time to pursue it, those cows who were minimalist fly hosts should be the mothers of future herd sires and their daughters should be high priority replacements.

I could go on and on with similar examples but I trust you to get the point. There are alternatives to the glossy ads through management changes and appropriate genetic selection.

You have to start parasite and pest management by knowing what you have, why you have it, and then evaluate your options for dealing with the little beasties.

The Basics:

- Environmental factors affect what pest and parasite problems you are likely to encounter and how severe a problem they will be.
- Length of the pest season is one factor.
- Parasites love warm weather.
- Rotational grazing does not lower parasitism.
- High grass residual reduces parasite problems.
- Genetics can lower parasitism.
- The greater preponderance of endophyte-infected tall fescue in the South makes cattle more susceptible to secondary stresses like parasites.
- Grazing management affects pest and parasite management.
- Long term rotations that change livestock species use can break parasite cycles, but just rotating cattle every 30 days through ten paddocks on a hundred acres does very little to reduce a parasite or pest problems

Think About Your Farm or Ranch:

- Make a realistic assessment of what are likely to be problems on your place with your livestock. What is the cause?
- Are flies an economic problem on your herd in your environment?
- Identify and cull those animals that get plagued by parasites.

38 A Shady Answer to a Shady Question

Tolerances are largely determined by what we have gotten used to.

Do cattle need shade?

Someone always has to ask the question at a workshop or field day. Seems like a pretty straightforward question. Kind of like, "What's the meaning of life?"

I don't really know what the answer is to either one of those questions, but I think they're both pretty complex questions.

Some cattlemen feel shade is an absolute essential for the well being of their cattle and make sure all of their pastures have plenty of shade available. Other ranchers having cattle in the open expanses of the Plains often point out those cattle have never seen a shade tree their entire life. That might be a bit of an exaggeration but comes pretty close to the truth.

At the Forage Systems Research Center in Linneus, Missouri, very few of the research pastures had any shade available, and there were cows there that went several years never seeing shade. All of this makes me think one of the first issues is, what are the cattle used to?

As with almost all aspects of life, our tolerances are largely determined by what we have gotten used to. Elsewhere we talked about cows that know they have to rustle their grub through the snow will do so, while cows that are used to having hay in front of them all the time need hay.

Far too many times to count, I have seen cattle bunched up under a shade tree when the temperature was less than 70 degrees. Are those cows realizing a benefit from shade or are they just indulging in one more bad habit?

Research on the shade needs of grazing livestock is about as sparse as shade in the Texas panhandle. The benefit from shade for feedlot cattle and confinement dairy cows has been shown, in some environments. But research evaluating grazing beef animal response to shade is scarce.

In the last several years, work at the University of Kentucky and the University of Missouri has shown physiological and performance response to providing shade for cattle grazing endophyte-infected tall fescue. However, the researchers in Kentucky questioned the economic value of providing artificial shade. Response to shade on non-fescue pastures has been much more difficult to demonstrate.

I think we have to first ask, why should cattle need shade?

Aside from the matter of an acquired habit mentioned above, why should cattle need shade? The obvious answer is to stay cool. But does being under a shade tree with 60 other animals really help lower body temperature?

That raises another question, is some shade better than other shade?

Natural shade has been shown to be more beneficial than artificial shade, in some environments. The evaporative cooling that occurs beneath a tree and the filtering of sunlight has a different effect than what occurs beneath a metal roof.

Research at Mississippi State University in the early 1950s showed an animal response to natural shade, but not to artificial shade.

Is a grove of trees more beneficial than a single tree?

Intuition would tell us that animals could disperse more

in a grove and avoid the tight bunching forced by lone trees. I suspect the thermal mass of 60 animals under a single tree is probably more detrimental to heat-stressed individuals than having no shade. The dispersed shade of a savanna environment is probably the ideal situation.

The broad question is, do animals need shade? The real question becomes, do certain animals need shade? And if they do, is it needed at certain times?

For about 30 years, the main research herd at FSRC calved from about February 15 to April 15, which placed the breeding season from about May 7 to July 7 until the mid-1990s when we went to a 45-day calving season with the same beginning date but ending 15 days earlier.

Initially they were red cows but most turned black over the years. Year in and year out those cows averaged 93% conception rate. And remember, there was no shade in those pastures.

From the late 1980s until 2002, my cows at home calved from April 1 to May 31, which put the breeding season from June 20 to August 20, right smack in the middle of heat, humidity, and fescue. Year in and year out my cows also averaged about 93% conception rate. My cows were all red and sometimes had shade and sometimes they didn't.

In 1995 we shifted part of the FSRC herd to April 15 to June 1 calving with breeding beginning about July 7. My cows at home had been doing fine calving and breeding in those windows and I had no reason to believe the FSRC cows would not perform just as well. First year conception was 92%, but the cows had an additional 60 days between calving and breeding compared to the earlier winter-calving cows.

In the second year, conception rate dropped to 68% and in the third and final year of the study it was only 48%. It was the final year because we ran out of cows!

What a wreck. Some critics immediately came to the conclusion that calving outside of the "traditional" February-March window just doesn't work. My conclusion was that we had the wrong cattle for what we were trying to do. Now comes the interesting part.

Our repro guys didn't really think it was a cow problem, but rather a bull problem. So in the third and final year we semen checked bulls every two weeks. With each successive test we saw semen quality and quantity crashing. Cows were blood tested and observed to detect cycling. The cows had no problem cycling in the hot summer breeding season. It was the bulls that couldn't stand up to it. There may have also been some embryonic loss among the cows due to heat, but that was not assessed.

We still have the question, do cows need shade in a summer breeding season?

I still don't know whether they do or not.

Would shade have kept those black bulls fertile and gotten the cows bred? I don't know.

Would using a heat-tolerant breed of bull on those black cows have gotten them bred? I don't know. (Looks like a great opportunity for someone who wants to do shade research with breeding animals.)

There is very good research regarding the shade usage of heat-tolerant versus non-heat tolerant breeds of stocker cattle. Dr. Richard Browning at Tennessee State University has compared Senepol and Hereford steers and their use of shade in both infected tall fescue and orchardgrass pastures. Senepol steers used shade much less than Hereford steers on both types of pasture and performed equally well.

Senepol steers performed similarly on orchardgrass and infected tall fescue pastures suggesting that there is a strong genetic basis for fescue tolerance.

Hereford steers on orchardgrass gained over twice as much as their contemporaries on infected tall fescue. Shade was available to all of the cattle, but while Hereford steers on fescue spent about 45% of their time in the shade, Senepol steers spent only 7% of their time in the shade.

Did the Senepols really benefit from that 7% of the time

in the shade? I don't know. If Dr. Browning can keep getting funding and keep his research alive, maybe we'll know a little bit more in a few years.

So there you have it, not much of an answer. But at least I didn't say "it depends" until the end of the article.

The Basics:

- Natural shade has been shown to be more beneficial than artificial shade, in some environments.

Think About Your Farm or Ranch:

- Do your animals need shade?
- If so, when?
- What kind of shade can you provide?

39 Coping with and Minimizing Pasture Bloat

Fear of bloat costs the livestock industry many times more than bloat itself does.

One of the worst ways to start your day on the ranch is going out to the pasture and finding a half a dozen cows swollen up like blimps with their legs sticking up in the air.

Bloat is one of the greatest fears for most graziers. We all have heard the horror stories from the old timers at the feed store about their neighbor who lost a third of his herd one morning. Veterinarians love nothing more than to show big bloated corpses on pasture right after an evening dinner meeting.

While an individual producer may be devastated by a bloat occurrence, the overall livestock industry actually experiences losses due to bloat of less than one half of a percent annually.

What exactly is bloat?

Understanding basic rumen physiology and function is essential to understanding bloat. The great physiological advantage ruminants possess is that great big fermentation vat known as the rumen. Microbes in the rumen are able to take plant fiber that is indigestible by monogastrics and extract energy for use by the ruminant animal. All fermentation processes generate gases including carbon dioxide and methane. Belching is part of normal ruminant function and allows gases generated in the rumen to be expelled. If gas were not expelled on a regular basis, the rumen would swell up putting ever increasing pressure on the lungs and other internal organs. Eventually the windpipe shuts off and the animal dies of suffocation. In other words, the animal bloats.

Within the rumen is a great deal of fluid mixed with partially digested plant material. Atop the fluid is a foamy mat of less digested forage and above the mat is the gas pocket which is regularly evacuated through belching.

Certain types of plants cause the foamy portion of the rumen contents to expand and occupy more and more space. Eventually the foam layer can cover the opening to the esophagus trapping the gas inside and starting the bloating process. When it occurs, it can happen very fast, which is why most producers who encounter bloat find dead animals and not live ones. The condition where bloat is most likely to occur is on legume-dominant pastures in a vegetative state. Not all legumes cause bloat, nor is the problem exclusive to legumes.

Extremely high quality grass pasture with crude protein level in excess of 30% have on rare occasions been reported to cause bloat. By far and away, most bloat occurs on pastures containing either white clover or alfalfa. Red and sweet clover can also cause bloat, but at lesser frequency.

Birdsfoot trefoil, lespedeza, and sainfoin are some of the legumes that don't cause bloat. Legumes high in the chemical saponin seem more likely to cause bloat while legumes containing tannin compounds do not cause bloat. Tannins are known to bind with proteins and allow them to pass through the rumen undigested.

So how can we minimize bloat risk in our operation?

There are several common recommendations to minimize bloat risk. None of these practices guarantee that bloat will not occur, they only lower the risk. The highest risk situations are grazing pure stands of bloat-causing legumes. Sudden changes in the diet are likely to induce bloating. If you are practicing rotational grazing and using grazing periods longer than two or three days, the change in diet composition animals experience when leaving one paddock and going to the next can trigger bloating.

When livestock are moved to a new paddock they begin by selectively grazing the highest quality plant material available. In the case of legumes this can easily be in excess of 30% crude protein. This high protein level can result in excess froth development in the rumen and bloat can develop rapidly.

If the animal also ingests a large quantity of stem simultaneously, protein content is lowered and more fiber is added to the rumen mat, thus reducing the risk of bloat. Single-day grazing periods reduce the degree of selectivity in grazing resulting in greater stem consumption.

The first idea is to keep the diet as consistent as possible and avoid moving animals when they are hungry. Allowing the legume to advance to a more mature stage before grazing can also reduce bloat risk. As plants mature, leaf:stem ratio drops as does the protein level. Stemmier legumes are less risky than leafier legumes.

Feeding grass hay or some other low protein, high fiber supplement lowers bloat risk just by dilution effect. The challenge is getting animals to eat a low quality dry forage when they have the alternative of lush pasture. Some producers report that their animals seem to have the good sense to eat the dry material to balance their diet. Not everyone has the good fortune of owning such wise animals.

If we take a little more proactive approach, we can avoid the high risk situation of pure stands of legumes. Planting a grass-legume mixture significantly reduces bloat risk. Once again it is the dilution factor reducing protein level and raising fiber content. Even with grass-legume mixtures, short grazing periods are helpful for minimizing the opportunity for selective grazing.

On either pure stands or mixtures, always avoid grazing

too short as this puts the animals into a hungry state when moved to a new paddock. The combination of voracious appetite with the opportunity to selectively graze high-risk legumes is the most common cause of bloat.

Bloat is usually a manageable problem, but even with the best of management, bloat can strike and be disastrous for an individual producer. Not using legumes for fear of bloat results in higher production costs and lowered performance. Overall though, fear of bloat costs the livestock industry many times more than bloat itself does.

The Basics:

- Understanding basic rumen physiology and function is essential to understanding bloat.
- Certain types of plants cause the foamy portion of the rumen contents to expand and occupy more and more space.
- The conditions where bloat is most likely to occur is on legume-dominant pastures in a vegetative state.
- Not all legumes cause bloat, nor is the problem exclusive to legumes.
- The highest risk situations are grazing pure stands of bloat-causing legumes.
- Sudden changes in the diet are likely to induce bloating.
- Keep the diet as consistent as possible and avoid moving animals when they are hungry.
- Allowing the legume to advance to a more mature stage before grazing can also reduce bloat risk.
- Feeding grass hay or some other low protein, high fiber supplement lowers bloat risk.
- Planting a grass-legume mixture significantly reduces bloat risk.
- Short grazing periods are helpful for minimizing the opportunity for selective grazing.

Think About Your Farm or Ranch:

- How can you minimize bloat risk in your operation?

Fencing

40 Get the Right Energizer for the Job

Getting the right one for the job and installing it properly are the first steps to creating an effective, low-maintenance fence system.

The heart and soul of a reliable electric fence system is the energizer. Getting the right one for the job and installing it properly are the first steps to creating an effective, low-maintenance fence system. With all the options available it can be challenging to make the right choice.

Energy capacity and power source are the main decisions to make. After that everything else is bells and whistles. Manufacturers have tried to express fence power a lot of different ways over the years. When you go shopping you can find energizers advertised as powering 10, 20, or even 200 miles of fence. Others might indicate a particular energizer to be suitable for 40 acres and another for 400 acres.

Comparisons like this make certain assumptions about quality and condition of the fence that may or may not be relevant to your situation. Within a particular brand of fence energizers, these can be valid comparisons because the manufacturer, hopefully, has made the same assumptions for all of their units. However, they have very little relevance when trying to compare brands.

Energizers can be compared relatively objectively on the

basis of electrical output. The most common unit of energy used for energizer rating is the joule. One joule is an electrical output of one watt per second. Because the electrical pulse emitted by the energizer is so short, the joule rating is still a little nebulous.

Modern fence energizers are described as being low-impedance. This basically means there is little internal resistance to energy release. To be classified as a low-impedance energizer, the duration of the electrical pulse has to be less than .0003 seconds, which isn't a very long time.

Old-fashioned fence chargers may have had energy pulses up to one tenth of a second. The difference between pulse duration is the modern energizer produces a sharp spike of energy that flows with greater amperage than older energizers. The actual voltage output level in modern energizers is generally lower than older models. It is the sharp spike of energy at higher amperage that makes them kick so much harder. The narrow base of the spike is what prevents the fence from shorting out as easily as older energizers.

I always thought physics was a fairly precise science, so it would seem that rating energizers on an absolute electrical unit would really clear the haze around energizer capacity. It helps but isn't perfect. Slight differences in pulse length can make a big difference in joule rating. Still it is the best comparison available and we can start relating joule output to our potential needs.

How many joules does it take to power a mile of fence?

That's a good question and it gets the usual "it depends" for an answer. Pick up literature from one fence company and it says to figure six miles per joule. But you can get literature from another company and it says only one-half mile per joule. Is a joule from the first company 12 times better than a joule from the second company? No, they just use different sets of assumptions.

The first is thinking about perfectly constructed, single-wire fence with no vegetation load while the second outfit is thinking multi-wire sheep fence with a ton of vegetation on the fence. In other words, best case versus worst case scenario. Your situation will probably be somewhere in between.

What I have concluded based on my experience with over 40 miles of electric fence at FSRC and 25+ miles at home is that figuring one joule per mile of fence will generally ensure that you always have adequate power for livestock control. I am assuming that some of your fence will be multi-wire perimeter or lot fence and some will be single wire paddock fencing. Figure up the total of miles of fence expected and get energizer capacity to match that number.

If you plan on very little multi-wire fence and will be running primarily single-wire internal subdivision fence, then you can get by allowing more miles per joule. I would not go beyond three miles per joule to be on the safe side.

You've decided joules are the criteria you're going to shop by and now you see the energizer has two different joule ratings: output and stored joules. Output is what the unit dumps into the fence on every click. Stored energy means the unit has the capacity to increase output in response to load on the fence. In other words, it is backup energy. Make your purchase based on output energy, not stored energy.

Do I run all of my 25 miles on a single energizer? No, we run three energizers to power different parts of the farm. Once you get more than five or six miles of fence, I would rather split the system and have multiple energizers. That way if one crashes, you have backup power already in place. One very critical point, though, is **NEVER** have two energizers hooked to the same fence. It will destroy one or both of the units. Always unhook one energizer before joining the systems.

Install a switch between the fence sections so that both sides can be powered by either energizer if need be. Put a hang tag on the switch that says something like, "Do not close this switch under penalty of death." You don't want some well meaning tourist or hired hand closing the switch to do you a favor.

Once you know what size energizer you need, decide what power source will work best for you. There are three basic choices: mains power or plug-in, battery, or battery with solar support. Cost follows this order with mains units being the least expensive per joule of output capacity.

Because mains units are least expensive and have the lowest maintenance requirements, they are my first choice. The electrical cost to operate even the biggest plug-in unit is just pennies per month so operating costs is a non-issue. I prefer to either build all-electric perimeter fence or run offset feeder wires on existing fences to deliver power to all parts of the farm or ranch. Plug-in units range in size from less than one joule to 30+ joules so you can always find one that fits your needs.

Battery units are available in a wide range of capacities but don't get quite as big as the biggest mains units. While I use mostly mains powers, the battery units that I find most useful are the little strip-grazing units that run on either a single 9-volt lantern battery or a series of "D" cell batteries. They have a good kick for temporary fences used to graze crop residue or the lawn or whatever. A set of batteries will last several weeks.

Some battery units use external wet-cell automotive batteries while others use either internal dry or gel cells. Some are rechargeable and some are not. For ease of use, I prefer the internal gel-cell units.

More and more of the newer battery units feature variable draw features. This allows the energizer to sense the energy draw on the fence and adjust energy output to demand. This makes batteries last a lot longer than they did with older models. Increasing amount of vegetation touching the fence is the primary cause of increased draw on the fence.

To avoid the need for changing or charging batteries manually, solar powered units or add-on solar panels are available. There are two basic types: Modular units that contain the energizer, battery, and solar panel all in a single case or free standing battery units with a solar panel as an accessory. We used several solar units of both types at FSRC for many years and found most of them to be very satisfactory.

Because they are considerably more expensive up front, I tend to recommend solar units only when mains power is not a viable option. Use the same criteria for determining size of the energizer as with mains power. A joule from a solar unit has the same energy potential as a joule from a mains unit. Source of power is the only difference.

Buying the right energizer for the job is just as critical as getting the right sized tractor for your farming needs. Energizers come in a lot of different colors just like tractors do. Basically they all have the capacity to get the job done, just make sure you get one that has good dealer support and that you can get parts for it when needed.

The Basics:

- Energizers can be compared relatively objectively on the basis of electrical output expressed as joules.
- One joule is an electrical output of one watt per second.
- Figuring one joule per mile of fence will generally ensure that you always have adequate power for livestock control.
- Make your purchase based on output energy, not stored energy.
- NEVER have two energizers hooked to the same fence. It will destroy one or both of the units. Always unhook one energizer before joining the systems.
- There are three basic choices for power sources: mains power or plug-in, battery, or battery with solar support.
- Increasing amount of vegetation touching the fence is the primary cause of increased draw on the fence.
- There are two basic types of solar units: Modular units that contain the energizer, battery, and solar panel all in a single case or free standing battery units with a solar panel as an accessory.

Think About Your Farm or Ranch:

- How many joules will it take to power a mile of fence for your operation?
- What power source will work best for you?

41 Grounding and Lightning Protection

To get the most out of your energizer, it needs to be properly grounded to absorb the returning energy flow.

Understanding how an electric fence works will help you do a better job of installing your entire fence system. It is really a simple process. The energizer sends a positive charge into the fence line, the soil is naturally negatively charged, when the animal comes in contact with both fence and soil it becomes the switch closing the circuit. The electrical energy flows through the animal and back to the energizer grounding system to complete the circuit. To get the most out of your energizer, it needs to be properly grounded to absorb the returning energy flow.

The energizer's ground system is essentially an antenna for receiving returning electrons. The bigger the antenna, the greater its capacity to absorb energy flow. The size of the antenna is defined by the volume of conductive soil it interfaces. Most manufacturers' literature emphasizes the need for multiple ground rods to create an effective grounding system. Recommendations vary a little from brand to brand but a general policy is install at least three foot of ground rod per joule of energizer output. On drier soils more ground rods will increase effectiveness of the fence.

The spacing of those ground rods is a critical factor. Electrical flow to the ground rod from the soil occurs at a rate determined by soil conductivity. If your energizer needs three ground rods, they need to be spaced well apart to interact with different soil volumes. Three rods driven in a tight cluster have little more effect than a single rod. Recommendations for spacing range from 10 to 25 feet. We have generally gone with 10- to 15-foot spacing with the rods attached in series.

Either copper or galvanized rods may be used for grounding. Copper has greater surface conductivity than does galvanized so has that advantage, but most leadout wire sold by fencing companies is galvanized so using galvanized rods avoids potential for conductivity loss due to the electrolysis that occurs when two foreign metals are joined. If you choose to use copper rods, use copper wire to connect the rods to the energizer. Most energizer terminals are stainless steel so there is no worry for electrolysis at that connection.

A lot of people have asked if galvanized pipe can be used instead of store-bought solid rods. Pipe is fine as long as the ground wire is firmly attached to the pipe. When using pipe for ground rods, I use stainless steel hose clamps for the attachment. Stainless steel does not react to galvanized rods and the clamps can be tightened to make very good electrical connections.

Ground rods are fine for places where you can drive them straight into the ground the full length of the rod, which is typically 6 to 8 feet. What about when you hit rock at eighteen inches? That is when you need to get creative with your ground system. Remember ground surface area and soil conductivity are the key factors to good grounding.

Locating ground rods where the soil is a little deeper and a little more moist can help in those shallow-to-bedrock situations. The ground system does not have to be immediately adjacent to the energizer location but needs to be directly wired to it. Locating the ground rods a couple hundred feet away in a ditch bottom where the soil is deeper and stays moist longer can work. In soils with moderate depth to bedrock, just drive the ground rods in at an angle rather than straight down.

Another method that works is to backhoe down to near rock level and lay a sheet of galvanized roof metal in the bottom of the trench. Firmly attach the ground wire to the roof metal with a stainless steel or galvanized nut and bolt. Refill the trench with a heavier soil or a layer of bentonite clay to increase water retention. The ideal location for this type of ground is on the north side of a shed without gutters. The north side shade and water runoff will help the ground stay more conductive through drier times of the year.

Galvanized road culverts make excellent fence grounds due to their large surface area and they are generally wet on the lower side. Just use a nut and bolt to attach the wire to the culvert. Many people are tempted to use well casing for grounds. While they do make excellent grounds, it is an unsafe practice. If the fence is dead shorted somewhere, the water in the well becomes charged and can create real problems for livestock or anyone else trying to use the water. Also avoid using any existing ground system that serves any other purpose.

The other aspect of energizer installation important to consider is lightning and power surge protection. For every energizer damaged by lightning about five more are damaged by power surges on the input side. A number of factors cause power surges to occur, particularly in farm electric systems. If your farm is located at the end of the transmission line, you are more likely to experience power surges. Weather related electrical outages are often followed by power surges when the power comes back on.

Probably the biggest culprit for power surges is inadequate capacity in the service entrance. Many older barns and sheds have woefully inadequate service boxes. As more and more electrical demands have been added on the farm, fuse or circuit box capacity has not kept up. If your shop lights blink every time you strike an arc with the welder or turn on your table saw, you have a good chance of blowing your energizer. The quickest protection is to simply use a plug-in surge protector wherever you install a fence energizer. In the long term, plan to upgrade your service capacity for the safety of both you and your energizer.

On the fence side of the energizer to protect from lightning, I like to use a two-stage lightning protection system. The first stage is a lightning diverter located where the leadout wire from the energizer joins the fence. There are several types of diverters with the porcelain twin tower being the most common type. All diverters feature two leads that are spaced far enough apart that normal fence voltage won't arc the gap but the high voltage lightning surge will. The top terminal attaches to the fence wire while the bottom terminal goes to a separate ground system from the energizer.

Lightning is always seeking the path of least resistance to return to ground. If you have done a good job of grounding your energizer, that could easily be the path of least resistance. You want to provide lightning with a better opportunity to ground before it hits the energizer; therefore, you want your lightning diverter to have a better ground than the energizer.

The second stage is an induction coil located between the energizer and the diverter. An induction coil is a loop of insulated leadout wire with about six to ten loops taped together to form a tight circle about ten inches in diameter. When the lightning surge hits the loop, it encounters resistance and reverses its flow back towards the lightning diverter ground. For about the first ten years FSRC used low-impedance energizers, we used only the lightning diverters and still had a number of energizers damaged by lightning. Once we began using induction coils in combination with diverters, we never had another energizer damaged.

I can remember a couple of occasions at FSRC when someone from the farm crew would come in after a storm and tell me we had lost an energizer. I asked if that location had an induction coil and the answer was no. Eventually all installations had induction coils installed and our energizer losses ended. To get the most out of your energizer, pay attention to the installation instructions and make sure you get it adequately grounded. Then protect your investment with a good two-stage lightning protection system. Most manufacturers' warranties are void if you don't use their lightning protection system, so do it right the first time.

The Basics:

- The energizer's ground system is essentially an antenna for receiving returning electrons.
- A general policy is to install at least three feet of ground rod per joule of energizer output.
- The spacing of those ground rods is a critical factor. Rods should be at least 10 feet apart.
- Galvanized pipe is fine as long as the ground wire is firmly attached to the pipe.
- Ground surface area and soil conductivity are the key factors to good grounding.
- Galvanized road culverts make excellent fence grounds when rods won't work.
- Never use well casings for grounding.
- Use a plug-in surge protector wherever you install a fence energizer.
- Your lightning diverter should have a better ground than the energizer.
- Using induction coils in combination with diverters helps prevent damage to energizers.

Think About Your Farm or Ranch:

- What will you use for fence grounds?
- Based on your soils, where and how will you locate your ground rods?
- Are you protected against power surges?
- Are you using induction coils and lightning diverters?
- Plan to upgrade your service capacity for safety.

42 Perimeter Fences for Your Grazing System

Properly constructed and energized, electric fencing will contain more creatures than will traditional fencing.

Perimeter fences are an important part of your grazing system as they serve two primary purposes. 1) They keep your animals in and you neighbor's out. 2) They provide a distribution network for fence power.

Many people still think in terms of a physical barrier when it comes to perimeter fencing. Even though they readily accept electric fencing for interior fences, they still feel the need for a perimeter fence to be much more substantial. Wooden pole or board fence, woven wire, or barb wire are common forms of traditional perimeter fencing. Many people have a greater feeling of security knowing at least their perimeter fence will hold anything. Of course, properly constructed and energized electric fencing will contain more creatures than will traditional fencing.

One good reason some people still install the heavy-duty, physical barrier is because state law or insurance policy dictates the use of traditional fence. Some localities do allow electric fencing for perimeter fencing but they have very specific requirements for fence construction. The published requirements for legal electric fencing that I have seen from several different states have all required a minimum of four electrified wires. Even though electric fencing makes economic and security sense for boundaries and road frontage, be sure to check the legal ramifications of such a fence in your specific location. General considerations aren't good enough. For example in Missouri, legal fencing is determined on a county by county basis, so what may be legal for your next door neighbor may not be legal for you if the county line is your boundary.

Insurance requirements are an important consideration anytime you are dealing with livestock. In most states, if a passing motorist hits one of your livestock on the highway, your legal liabilities are affected by the status of your road frontage fence. If your fence does not meet published legal standards, your liability is much greater than if you had a legal fence containing the livestock.

Once security and legalities are taken care of, function of the fence becomes the consideration. I think one of the greatest advantages of electric perimeter fences is that they provide a framework for delivering electrical power for fences all across the farm. The need for several battery energizers for different parts of the farm disappears when a hot wire encloses the entire property.

When constructing a new fence, making it all electric is easy to do. A 5-wire electrified hi-tensile perimeter fence typically costs about 60% of a 5-strand barb wire fence. For all classes of cattle, 4-wire is adequate. For the equivalent security of a 48-inch woven wire fence, we usually install an 8-strand hi-tensile fence configured as a hot-ground fence. A hot-ground fence has four wires hot and four wires grounded, with the bottom wire being the first ground wire and alternating with hot wires from there on up.

For a hot-ground fence to be most effective, the ground wires should be linked and connected to the ground terminal of the energizer. With this wire arrangement, anytime an animal contacts two adjacent wires, one hot and one ground, the full grounding system of the energizer is directly brought to bear on the culprit animal. This works particularly effectively with sheep and hair goats whose fleeces tend to insulate them from shock. It also makes an effective predator deterrent for coyotes and dogs that tend to climb fences.

You can also build a combination perimeter fence that incorporates the best of both worlds. A fence arrangement that we have used to satisfy the desire of neighbors who still want physical containment while incorporating power distribution for our use is a 5-wire fence consisting of four barb wires and one electrified hi-tensile smooth wire. The hot wire is usually the middle wire or second from top. Another advantage to this fence over a conventional barb wire is greatly increased longevity. It is not sun, wind, and rain that causes fence to wear out prematurely. It is animal pressure, even on barb wire. Replacing one barb wire with the one hot wire in a 5-strand fence effectively doubles the life expectancy of the fence by deterring animals from rubbing or challenging the fence.

For woven wire or wooden fences, add one hot wire on either offset brackets or wood post insulators on the side of the fence receiving animal pressure. If both sides receive animal pressure, put a wire on each side. I cannot begin to count the number of farms where I have seen very expensive woven wire fences nearly demolished within a few years because of excess animal pressure. Protect your fence investment with a little bit of electrical power.

One of the challenges in an extensive electric fence system is getting power across gates and other obstacles. Carrier lines can either be run overhead, underground or directly across. Going overhead is fairly easy and can be accomplished with extra tall wooden posts, lumber, fiberglass, or plastic pylons. Whatever material can be obtained cheaply is usually what gets used. My personal preference is fiberglass suckerod because of its durability and flexibility. Drilling holes through the rod near the top and threading the wire through eliminates the need for any insulators or other attachments. If you can get it directly from oil field suppliers, the suckerod comes in 36-foot joints in 1 to 1 1/4 inch diameters. Cutting a joint in half yields two 18-foot pieces, which gives adequate clearance for just about anything to get under it.

When going overhead, I prefer to use aluminum wire due to its lighter weight and greater conductivity. One concern about going overhead is lightning attraction. If you live in a lightning- prone area, it is a good idea to install a lightning diverter on the energizer side of every overhead.

Putting your crossings underground gets them away from lightning risk and possibly catching on some piece of equipment passing beneath. Plus it makes the farm or ranch more attractive. If going underground, use only wire made especially for the purpose or enclose a bare wire in plastic pipe or conduit. Double-insulated leadout wire is made for underground use but still sometimes fails or has flaws in the insulation leading to an underground dead-short, which is the biggest pain in the neck in the whole electric fencing arena.

The wire does not have to be buried particularly deep unless it is a gate that receives a lot of vehicle traffic during wet weather. I generally just dig a trench the width and depth of a garden spade or sharpshooter, as they are known in some parts of the world.

To do the job right and minimize any risk of failure this is how I do it. Use double-insulated leadout wire and run it through a plastic pipe. Either polyethylene or PVC work fine, but PVC is more resistant to any potential rodent damage. Seal the ends of the pipe where the wire emerges with high-quality silicon sealer to prevent any water from entering the pipe. Another method to prevent water invasion is to use elbows and have the wire emerge from the pipe on the down side. By the way, it is ten times easier to build the pipe assembly around the wire than to try to feed the wire through a pre-constructed pipe assembly.

For the folks who like to keep things really basic, just run a wire gate along your main gate on the animal pressure side. You generally have to get out to open the gate anyway so you might as well open a wire gate also. This is the cheapest and easiest way to carry fence power past your gates. Just make sure to install the wire gate so that it is dead when you unhook it. You can hang the gate handle on the swinging gate and it is out of your way. When you come back out, just remember to hook the wire gate so you have power on the other side. This also keeps bulls from crushing your pipe gates along the highway.

A perimeter fence needs to be able to hold your livestock where they need to be and be recognized by the authorities as a legal fence. After that it is largely your preference. Building electrical power into the fence provides benefits all across the farm or ranch.

The Basics:

- Electric perimeter fences provide a framework for delivering electrical power for fences all across the farm.
- A hot-ground fence has alternating hot and ground wires, with the bottom wire being the first ground wire.
- For a hot-ground fence to be most effective, the ground wires should be linked and connected to the ground terminal of the energizer.
- Replacing one barb wire with the one hot wire in a 5-strand fence effectively doubles the life expectancy of the fence by deterring animals from rubbing or challenging the fence.

Think About Your Farm or Ranch:

- Does your perimeter fence meet local legal and insurance requirements?
- What kind of fence arrangement will work best for your livestock species?
- Does your perimeter fence need to exclude predators?
- How will you run your carrier lines across gates and other obstacles?

43 Permanent Subdivisions — Planning & Using

Subdivision fencing is a management tool to help accomplish your goals.

Subdividing pastures into smaller units is the beginning of gaining management control of your forage resources. How much you subdivide is a question of how much control you want. Initial subdivision should break the landscape into manageable units based on size, location, composition, and productivity. Subdivision fencing is a tool to help accomplish those goals.

In the past, subdividing pastures was a major challenge due to the cost and labor requirements for constructing fence. One of the really picturesque features of Ireland and the UK are the many miles of stone fence that subdivide the countryside. While a boon to tourists, the functionality and mobility of those fences is somewhat limited. Even 5-strand barb wire, which is a much cheaper alternative to wood and stone fences, is a fairly high price option for subdividing pastures to anything more than quarter sections.

Modern electric fencing is what has made subdividing pastures for intensive management possible. When first exposed to electric fencing, most people still fall back on their conventional fence paradigms. I was guilty as anyone of that and it takes some experience to get past those barriers. In 1981 we needed to install some new fence at FSRC to divide 20-acre pastures into 10-acre sections. All previous fencing had been done as 5-strand barb. We decided to be daring and try just four smooth hi-tensile wires with 20-foot post spacing. That worked fine so we did some 3-wire fence the following year. Soon we found ourselves taking a wire off the four wire fence to use somewhere else because it was entirely unnecessary.

After several years of using electric fence, we became confident enough to use just single wires to separate different research groups of animals and eventually evolved to enough confidence in the fence to use only a single strand of polywire to separate research treatments. Our herdsman found he was more confident in a single hot wire to keep bulls separated than in 5-strand barb. I emphasize that it was an evolutionary process. We had to overcome existing paradigms and gain confidence in new technology. Over the years I've noticed that novice graziers who had no previous experience in the livestock business learn to use minimal fencing faster than veteran cattlemen, because they have no paradigms to overcome. Unlearning is often a more difficult task than learning.

Where I have ended up with permanent subdivision fencing for cattle, bison, and horses is a single strand of 12.5 gauge hi-tensile wire on solid corners with line posts the deer can't knock the wires off. In dry environments, whether that be due to low rainfall or soil conductivity, two wires are needed for effective animal control. One as a hot wire and one as a ground wire directly connected back to the ground system of the energizer. For sheep and goats, three wires is my choice for permanent subdivision. In wet environments, make them all hot and, in dry environments, make the middle wire ground. I like to see sheep fences kept considerably tighter than cattle fences and post spacing kept fairly close. This forces the wire through wool and hair coverings if the sheep or goat tries to push through the fence. I have seen full-fleeced sheep walk unconcernedly through 3-wire fences carrying 7000 volts because the loose wires just rode over top of their wool. We started building 3-wire fences with 40 to 50 ft post spacings but reduced them down to 20 to 25 ft to keep the wires tauter.

Corners can be constructed from a wide range of materials from living trees to PVC. What you choose to use depends on availability and how much bearing strength is required. This is one area where I have stayed pretty backwards. I still prefer to set a wooden corner. Not just any wooden corner, but a hedge post. Known as hedge apple, osage orange, or bois darc (*Maclura pomifera*) in different parts of the country, this is the most durable wood in North America. Fortunately for us, our farm still had a hedgerow on almost every 40 line as well as numerous groves in the abandoned areas. In our environment, a 5-inch hedge post lasts two to three times as long as a 10-inch CCA treated pine post. A 10-inch hedge post is a monument for your grandchildren to remember you by.

Other options include heavy fiberglass or braced fiberglass suckerods, steel T-posts with bracing hardware, PVC pipe either filled with concrete or just capped off, recycled plastic posts, and a whole array of other stuff. As long as it can solidly hold the end of the wire and be insulated, it can work for an end post. Of the above options, I have found the heavy fiberglass to be most satisfactory. The posts we get are reject quality from a local manufacturing plant and range in size from 1 3/4 to 3 inch diameter. Look for opportunities in your neighborhood for usable salvage material that can be bought cheap.

Single wire fences require minimal bracing. A post or pipe set deeper in the ground than the height of wire above ground is often enough in clay soils. A simple dead-man or bedlog brace can provide additional support to any of the post or pipe corners (*see diagram*). Depth of the corner is more important than diameter when it comes to holding in place.

For 3-wire fences, a little more bracing is needed. Most of the steel T-post corner assemblies will easily hold a 3-wire fence in place. On wood or large fiberglass corners, we use a floating knee-brace assembly *(see diagram)*.

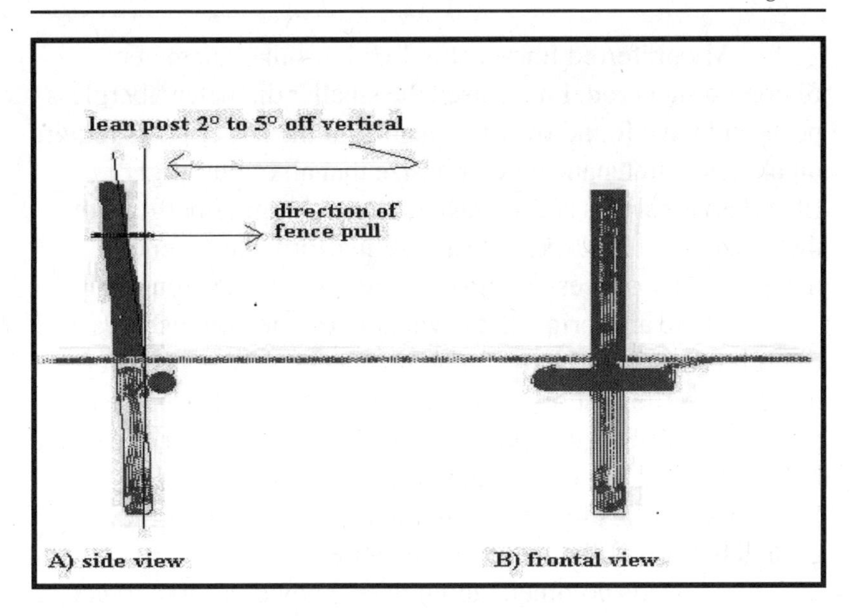

Dead Man Brace

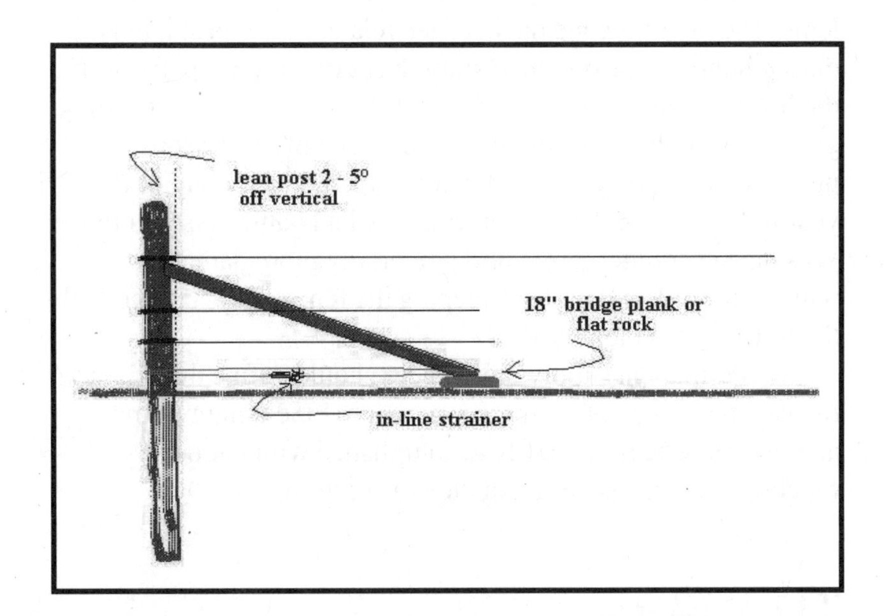

Floating Brace

My preferred line-post is 1 or 1 1/4 inch diameter fiberglass suckerod. I have used the smaller diameter fiberglass posts and have found over the years that the larger posts require much less maintenance. I emphasize that all of our larger fiberglass is surplus and salvage material. Virgin fiberglass of these sizes is quite expensive and is unaffordable as fence material. Most successful graziers are successful scroungers.

There are spring clips available for the various sizes of fiberglass posts but I have found that deer easily knock off or loosen even the best ones so I only use them on the smallest posts. For all fiberglass posts above 3/4 inch diameter, we drill a hole and attach the wire to the post with a wire cotter key. When deer hit this fence they might pull a post out of the ground but it is still attached to the wire.

As a side comment about deer. They don't like to get shocked any more than cattle do. Once your resident deer become accustomed to your fences they give very little problems. The only time we have a deer-related fence problem is during hunting season when stray deer cross the property and the local population gets panicked. Fifty weeks of the year they give us virtually no problem. Deer are one of the reasons I prefer to keep every foot of fence energized every day of the year. It keeps them from acquiring any bad habits. As you move West and encounter larger and potentially more damaging wildlife like elk and moose, keeping the fences fully charged all the time is even more critical.

Permanent subdivision fences should define the primary management units of your farm or ranch. Fine tuning of management may be more easily accomplished with temporary fencing which is discussed in the next section.

The Basics:

- Single wires can be used to separate different groups of animals.

- For sheep and goats, three wires is a good choice for permanent subdivision.
- Keeping the fences fully charged deters deer and other wildlife.

Think About Your Farm or Ranch:

- What materials are readily available in your area for end and corner posts?

44 When and How to Use Portable Fencing

Using portable fence is one of the easiest ways to fine tune management

A few years ago I calculated the potential hourly wage for using portable fence for strip grazing dry beef cows on stockpiled pasture in the winter. We had found through several research projects at FSRC that the harvest efficiency for grazing stockpiled pasture could be increased significantly with strip grazing compared to just turning a bunch of cows out on a large pasture. While giving two weeks' forage supply at a time allowed us to harvest about 50% of the stock-piled pasture, 3-day strips yielded 70% utilization while a daily strip gave 80% utilization.

With permanent fences spaced about 400 feet apart, I can reel up a polytape and reset it in less than 10 minutes. Each day a cow can graze instead of eating hay saves us about 75¢. With a herd of 100 cows, that amounts to $75 saved each additional day they can graze. By strip grazing daily rather than just turning the cows out, every time you move the tape, you earn $75. If it takes 10 minutes to move the tape, each hour spent moving tape pays $450. Not a bad wage for farming. Really not a bad wage for any profession. Unfortunately you can't spend five 8-hour days every week moving fence.

The point is fine tuning grazing management can pay

back big dividends and using portable fence is one of the easiest ways to fine tune management. To effectively use temporary fencing you need to know what you are trying to accomplish and have the right tools to do the job efficiently. All of the things we have discussed before concerning animal requirements, plant growth response, intake, and residual are the things you are trying to accomplish. This section is about doing it efficiently.

We saw in the last chapter that permanent fence can be made much easier and cheaper by using the right materials. Moving temporary fence can be a daily hassle or it can be an enjoyable outing. It just depends on what kind of materials you are using.

I helped teach a grazing workshop in central Illinois several years ago in the heart of corn country, One of the producers we visited was a grass-based dairy farmer who had been using rotational grazing since the early 1950s. The farmer was an elderly gentleman who didn't believe in wasting anything. He had a well defined crop and pasture rotation system that had been in place for 40 years. He had beautiful, healthy soil. His pastures were alfalfa-orchardgrass. The farm had all the appearances of a truly sustainable operation.

Because of the crop-pasture rotation, he used no permanent fence other than his perimeter. Rather than using any of the modern fencing products, though, he was still using barb wire and rebar posts for his movable fencing. He was moving fences every other day. The most peculiar thing about his grazing system was that his strips were always parallelograms rather than rectangular as most people tend to do. Most people, especially Midwesterners, like their temporary fences to meet the permanent fences at right angles. I don't really know why. We just do.

I thought this particular farmer had discovered that perhaps animal flow was better with the strip fences meeting the lane at an obtuse angle. Everything else in his operation seemed to be so well planned, I figured there must be a very good reason for his unique strip design. I asked him about it and his reply was, "That's how long the wire was when I brought it out here the first time. I didn't want to cut it off and waste a short piece of wire."

He had been grazing in parallelogram paddocks for 40 years because that was the system he had started with and he was satisfied with the results. We all acquire habits and get used to using some particular product or method and we develop our prejudices around those experiences. I have my own biases when it comes to movable fence because I have done a lot of it and had my share of disappointments along with successes.

The first issue in movable fence is what do you use for the wire. Soft wire, cable, hard wire, aluminum wire, polywire, polyrope, or polytape are all available options. The full metal products offer superior conductivity but are burdened by weight and the need for at least minimal tools for installation. The polyproducts offer the advantage of lightweight and no tool installation, but may lack in conductivity in certain situations.

I do not use any metal products for temporary fencing. I have used soft steel wire, aluminum hi-tensile wire, and 16-gauge cable and found all of them to be a hassle. The way I deal with the lesser conductivity issue of polyproducts is I never ask them to do more than they are designed to do. The critical factor to efficiently using polyproducts in my book is building an appropriate permanent fence framework to use them in conjunction with.

I have previously emphasized spacing permanent fences at no more than 400 to 600 feet apart. At this distance, polyproducts are highly conductive and easy to use. While I have used polywire on many occasions with satisfactory results, my preferred temporary fence material is standard polytape. My number one reason for this is visibility. We have lots of deer and they easily see polytape. We move fence almost daily and livestock never seem to lose its location, even in cereal rye or sudangrass that got away from us. Personal prejudice, that's one advantage to being the author.

I have been on lots of field days and pasture walks where the host demonstrates their fence reels made from extension cord reels, welding wire spools, and anything else you can imagine rolling wire up on. Invariably they point out how much cheaper they are than the store bought reels made for handling polyproducts. Some of them work real slick and others you need to hold your mouth just right to make sure they work.

When I first started using polywire in the early 1980s, I didn't think I could afford the store bought reels and I wrapped polywire up on all kinds of things. Then I used a real reel and found that I could work so much faster the price of the store bought reel looked cheap. My position is if you use the tools every day, you should own the best tools. If you only use them occasionally, feel free to get by with something less.

Geared reels versus straight crank reels? I have both and really like the geared reels when I am in a hurry. If I'm not in a hurry, the 1:1 reels are fine. How often are you going to be in a hurry? Geared reels cost $10 to $15 more than standard reels. If you use them a lot, that cost difference is not much to make up.

There are some brands that crank a lot easier than others. Try several out at field days and trade shows and get something that feels good to you. Also look at how they hang from permanent fences. A reel that has double hooks for hanging will stay in place a lot better than a single-hook reel. Softer plastic is more durable that rigid plastic so check the flexibility of the spool and handles. If the plastic is hard and glassy looking, the reel is more likely to break down. We have reels we have been using for over 15 years with extreme satisfaction. I once bought six reels because they were cheap and every one was broken down within two years. Sometimes you really do get what you pay for.

For step-in posts, I prefer a post with a large step, narrow spike, and wire holders that are easy to use and are on the opposite side of the post from the step. Most posts have too large a spike to put in dry or frozen ground, have too small a step to get your foot on easily, and have the wire attachments on the same side as the step. A lot of posts get brittle in cold weather or after a couple years of use and start breaking down. Good posts will last ten years or more and can be used in all kinds of weather and soil conditions. No post can be put through a rock, though.

For ease of use, we tie a conducting gate handle onto the end to the polywire or tape so the temporary fence is energized as soon as it is hooked to the permanent electric fence. Some people prefer to use a non-conducting handle in conjunction with an alligator clip jumper to charge the fence. To me that is just one more gizmo you have to carry and one more step to setting up the fence.

That brings up the issue of working your temporary fences hot. If you have a properly insulated reel, step-in posts that easily hook and unhook from the fence, and has the step on the opposite side from the wires, it is safe and easy to work your fences hot. By working your fences hot you often save a trip or two across the field and you never forget about making sure the fence is energized when you leave the pasture.

In summary, there are lots of options available for temporary fencing. This is one area where you really seem to get what you pay for. There is a lot of cheap, low-quality junk out there. Look at what other people are using with satisfaction, shop around and try different products on a small scale first, then buy or build something that will work for you.The more you use the tools,the more you can afford to invest in them.

The Basics:

- Daily strip grazing cows on winter pasture pays back big dividends because the alternative is feeding hay at an additional 75 cents/day.
- To effectively use temporary fencing you need to know what you are trying to accomplish and have the right tools to do the job efficiently.

- The critical factor to efficiently using polyproducts is building an appropriate permanent fence framework with which to use them.
- If you use the tools every day, you should own the best tools.

Think About Your Farm or Ranch

- Is your permanent fence set up to make temporary fencing easy to do?
- What are you going to use for wire?
- What are the best tools you can afford for the right job?
- Can you use your portable fencing in all weather conditions?

Water Basics

45 Water Requirements for a MiG System

Water is second only to air as a critical need for survival.

When we think about the very fundamental needs for life, air and water come quickly to the top of the list. Luckily, we have air all around us and generally don't have to worry about providing it. I have been to a few places where air quality was a little questionable, but all in all it isn't too difficult to find a breath of air. Water can be a little different story.

Water is second only to air as a critical need for survival. Animal well being and productivity are more quickly affected by a water deficit than any other nutrient. Almost every metabolic function is water dependent. There is a basic maintenance water requirement and then an increasing water demand as productive functions increase. Animal species and breeds differ in their water requirements depending on their center of origin.

It is easy to understand that bigger animals have higher water demand than do smaller animals. Obviously a 1200-pound cow is going to drink more than a 120-pound ewe. But do ten 120-pound ewes drink the same amount of water as one 1200-pound cow? The answer is no, and the reason why goes back to points of origin.

Sheep and goats evolved in drier climates than did cattle. Because of low water availability in the native ranges, sheep and goats are more efficient in water use and in extracting water from the forage they consume. The dry pellets excreted by sheep and goats compared to the much wetter dung produced by cattle is a testimony to this difference in metabolic processes. Sheep have only 50% to 70% of the water requirement of cattle when compared as metabolic units.

Domestic cattle are divided into two separate species, *Bos taurus* and *Bos indicus*. The first group includes those breeds commonly referred to as English or Continental. These cattle developed in temperate climates of Europe and western Asia and are very tolerant of cold weather. The second group are the Zebu and similar breeds that developed in the much hotter environments of Africa and southern Asia. Zebu cattle had to develop multiple means of dissipating body heat in their extreme environments and are less dependent on water intake for metabolic cooling. Zebu may require 20% to 30% less water than English cattle when placed in the same environment.

The upper end of the thermal comfort zone for English type cattle is in the 60° to 70° F range. Water intake increases fairly linearly above this level at a rate of about four gallons for every 10° F increase in air temperature (*see figure*). Zebu cattle have a higher thermal comfort threshold and increase water consumption at a slower rate compared to English cattle.

Sheep and goat breeds exhibit similar patterns based on where they originated. The Navajo Chara sheep may go five to seven days without consuming free water and have little or no effect on performance while English breeds such as Suffolk or Dorset may need water about every three days to maintain performance.

Water demand is closely tied to feed consumption. Because water is necessary for digestion and just about everything that follows, as feed intake increases, water intake increases proportionally. One way to look at water demand is tying it to feed intake. If you are pushing performance and have

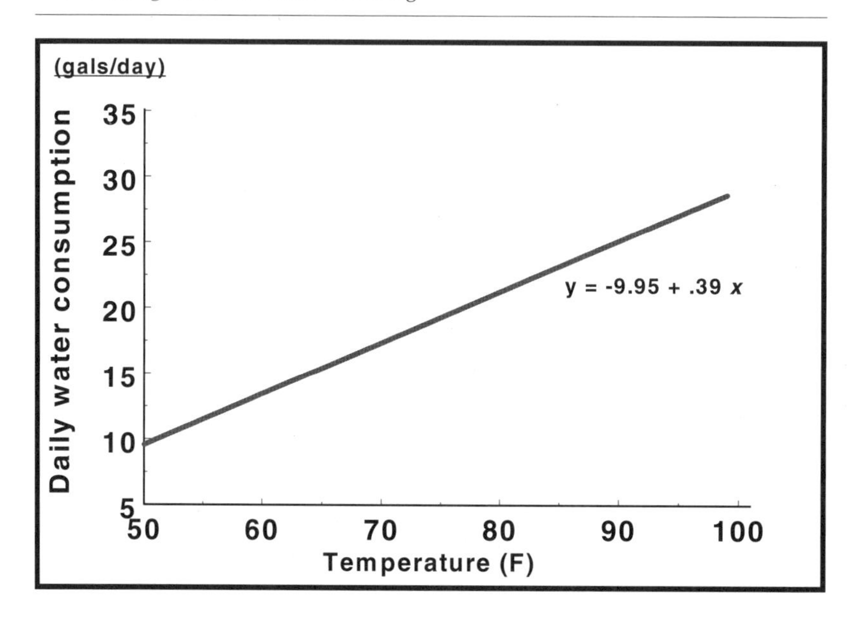

a good handle on the animal's energy intake, basic water requirement is fairly predictable. Additional water demand due to increasing air temperature must also be factored into the total requirement.

Because milk contains a lot of water and feed intake increases as lactation occurs, lactating animals have substantially higher water requirements than do dry animals. If you expect a cow's forage requirement to increase by 50% when she starts lactating, expect her water requirement to increase proportionally. There are lots of general water guidelines and tables published but very few take into consideration all of the factors affecting water intake.

Most guidelines say a beef cow will need between 15 and 20 gallons of water per day. When you're figuring what your cows will drink, you need to plan for variations due to temperature, stage of production, breed, and forage base. It is not uncommon for cows calving in April and May to be at peak lactation in early July. If she is a superior milking cow, grazing dry grass, and the air temperature is hovering around 95° F every day, she can easily be drinking 35 to 40 gallons per day. If your water system was designed to provide 20 gallons per/cow/ day, you have a real problem on your hands.

When planning the water system, figure the highest possible demand imaginable. Then, factor in another 25% water flow potential to account for pipe sedimentation and other system challenges that will occur over the next 20 years. If you do this, you can probably rest well even during the hottest days of the summer.

Quality of drinking water is another concern. Defining quality is a challenge in itself. Certainly the absence of toxic compounds is essential to quality. Nitrates, sulfates, copper, and a whole array of chemical and biological toxins may exist in water. For every toxin there is a threshold level for clinical symptoms, but what we frequently face are sub-clinical levels that may depress animal performance. If water contamination is a common problem in your area, have your water tested and determine if it is acceptable for livestock use. If it is not, you may need to develop some alternative water source.

Published research is mixed on effect of water quality on animal performance. Cattle seem to be the least selective about their drinking water and there is not a consistent response to improved water quality with beef cattle. Research has shown dairy cows respond positively to clean water. Sheep and goats are highly selective about water quality and will avoid drinking brackish water. On any given day, horses may choose to drink or not drink water of any given quality. They're just being horses.

Many stock water enthusiasts say you shouldn't expect livestock to drink water that you would not drink yourself. While that is a generous notion, only wells or springs are likely to provide water of that quality. Most cattle in the Midwest and South drink pond water of questionable quality. Installing a pipeline and tank system substantially increases water quality compared to drinking directly from ponds.

If your stock are experiencing chronic, low performance that can't be explained by any disease, parasite, or plant toxicity, have your water checked for a wide array of possible mineral or organic components that could be potential problems. Broad testing can be expensive, so try to narrow the targets down to those most likely to occur in your environment and management situation.

When you think you really understand animals and their drinking habits, they will do something to change your mind. The best you can do is make sure you're putting an adequate amount in front of them and trust them to do the right thing with it.

The Basics:

- Animal species and breeds differ in their water requirements depending on their center of origin.
- Sheep have only 50% to 70% the water requirement of cattle when compared as metabolic units.
- Water demand is closely tied to feed consumption.
- Lactating animals have substantially higher water requirements than do dry animals.
- When planning the water system, figure the highest possible demand imaginable. Then, factor in another 25% water flow potential.
- Installing a pipeline and tank system substantially increases water quality compared to drinking directly from ponds.

Think About Your Farm or Ranch:

- When figuring what your livestock will drink, plan for variations due to temperature, species, stage of production, breed, and forage base.
- Design or upgrade your water system to accommodate your livestock's peak seasonal needs.
- Have your water tested and determine if it is acceptable for livestock use.

46 How to Figure Water Demand for Your Stock

To figure water demand, you need to know a little bit about both daily requirements and animal behavior.

A water system needs to be able to deliver water to the livestock under the most extreme conditions and do it reliably. To figure water demand, you need to know a little bit about both daily requirements and animal behavior.

We've already discussed animal requirements and some of the factors affecting it. Behavior dictates how much water needs to be provided in what time frame. Watering behavior is affected by time of year, ambient temperature, water location, and the distance livestock must travel to reach water.

Water demand increases as ambient temperature increases so livestock visit the water tank more often in summer than they do in winter. Cattle may come to water only once per day during the winter and as many as 5 to 7 times daily in the summer, if water is close by. The farther they must travel to water, the longer they will hang around the watering point.

These behavior patterns are what determine how large a tank and what recharge rate are required. Ideally, the entire herd should be able to water in less than one hour even during the highest demand periods. If the herd can't water in this time frame, the likelihood of dominant animals in the herd leading the entire group away before everyone has had adequate water increases. Even though they are pathetically domesticated, cattle still have the basic instincts of wild animals and seek the shelter of the herd for protection. Cattle really prefer to maintain eye contact with other members of the herd. As long as they can do so, they are more likely to go to the water tank by themselves or with just a couple of buddies. Sheep and goats vary a lot in the herd tendencies depending on breed. Some are quite independent and may travel a good distance alone to reach water while some breeds exhibit extremely tight herd behavior.

On flat land, livestock can see quite a distance and they will travel independently to water a greater distance on level or open terrain compared to hill country. Cattle in hilly or wooded terrain will generally maintain herd behavior if they must travel more than a quarter-mile to water. Water demand increases as ambient temperature increases so livestock visit the water tank more often in summer than they do in winter.

When cattle come to water in warm periods, each individual will generally drink for two to four minutes at a rate of two to three gallons per minute. Each animal may drink between five to ten gallons per drinking event. If they are close to water and are making frequent visits, they may turn around and head back to pasture as soon as they and their companions have watered. If they are traveling a long way, they may take one drink and move away briefly and return for another drinking event. It's these habits and patterns that dictate how much tank space, reservoir capacity, and recharge that you must provide for each herd.

If a large herd comes to water all together, there should be adequate tank space for the entire herd to water within an hour. One rule of thumb is to figure 1 ½ inches of tank space per cow. A herd of 100 cows would require 150 inches or 12.5 ft of tank rim. This is approximately a four-foot diameter tank with access around the entire tank. About eight animals can drink at a time from a tank this size and if each animal drinks for four minutes, then it would take about 50 to 60 minutes for the entire herd to water. The problem with this scenario is that the tank only will hold about 175 gallons while the herd will need approximately 700 to 1000 gallons during the visit. To keep the tank from going dry, the pipeline system needs to recharge at a rate equal to the removal rate.

We can calculate required recharge by taking the 1000-gallon maximum requirement, subtract the 175 gallons in the tanks and divide by the time required for the herd to water (1000 gal-175 gal)/60 min=13.75 gal/min). In this example, the recharge rate needs to be at least 14 gallons/minute. That is more than a lot of existing livestock water systems can provide because they were installed with too small of a pipe. A well designed system with adequate pipe flow can easily handle this demand. This is why it is so important to determine the maximum water demand before you ever buy pump, pipe, or tanks.

For existing pipelines with low recharge rate, installing a larger tank is the best solution. In this situation, calculate the entire herd demand and plan for a tank with a capacity that will provide one-third the daily requirement. If we figure peak demand to be 30 gallons/cow/day, then our 100-cow herd needs 3000 gallons per day. So we need a tank that holds 1000 gallons. You can refer to capacity tables at the local farm and home store and find a suitable tank. If you have determined the recharge rate, you can reduce tank size by the amount of water the system can supply in one hour.

All this discussion has related to permanently located tanks. Many MiG systems rely on movable water tanks with small capacity that are easy to dump and move to a new location. Given the above discussion, it may be hard to believe that a hundred head of cows can be adequately watered with a 50-gallon tank. The key to making these systems work is breaking herd behavior and providing high recharge capacity.

The more intensive the pasture management, usually the smaller the paddock. This keeps livestock much closer to water and thereby encourages them to visit water more frequently and as individuals. As long as the tank can be refilled faster than the animals can drink it down, they won't run out of water. If three head can drink from the tank at a time, plan for at least 9 gallon/ minute recharge capacity.

I prefer a low profile tank in these situations to minimize the risk of animals pushing it around and turning it over. We always set the tank directly under a hot wire so animals can't get on all sides. Movable tanks should be checked every day to make sure they are working and that they are not tipped over. Any temporary tank that is tapered down to a smaller base should be avoided. They are opportunities for failure.

Sheep and goats drink at a much slower rate than do cattle and have lower requirements to begin with. Water demand is lower than in a cattle system so there is greater flexibility in material use and pipe sizing.

A final thing to remember is to figure the entire demand on the system. If you are running multiple herds, they all place a demand on the supply system. While it is critical to ensure that each watering point can meet the demand of the herd at that specific site, the pipeline system needs to maintain the required flow rate if every herd decides to get a drink simultaneously.

One weakness in the first system I installed on our farm was that it was a dead-end system. If one herd closer to the pump was drinking, flow rate at a more distant point serving a second herd was significantly reduced. This problem can be minimized by installing the pipeline as a full loop system. That is, tee the line at the pump and pump water both directions in the loop. This maintains more constant pressure throughout the entire system and keeps even flow at every water point.

The Basics:

- Behavior dictates how much water needs to be provided in what time frame.
- Water demand increases with ambient temperature, so livestock visit the water tank more often in summer than winter.

- Water can be adequately supplied by either providing a large drinking reservoir or in a small tank combined with rapid refill rate.
- The key to making portable water systems work is breaking herd behavior and providing high recharge capacity.
- Movable tanks should be checked every day to make sure they are working and that they are not tipped over.
- Install the pipeline as a full loop system — tee the line at the pump and pump water both directions.

Think About Your Farm or Ranch:

- What are the optimum locations for water points in your paddocks?
- One rule of thumb is to figure 1 ½ inches of tank space per cow. A herd of 100 cows would require 150 inches or 12.5 ft of tank rim. Are your water tanks and pipelines sufficient for your herds' demands?
- For existing pipelines with low recharge rate, calculate the entire herd demand and plan for a tank with capacity that will provide one-third the daily requirement.
- Figure the entire demand on the system, especially critical with multiple herds or species.

47 Water Sources & Options for Your System

Whatever water source you choose to use, its reliability is a critical factor.

Livestock require water and can utilize it from several sources. The most obvious one is the stock tank, which might draw its water from a number of possible supplies. Wells, springs, streams, ponds and lakes, dugouts, and public water are all possible water sources.

In addition, water contained in feed and forage, dew on standing pasture, and snow are also sources used by livestock. The latter are generally figured to be fairly minor sources by most authorities, but at times they can provide the full water requirement of the animal.

Fresh pasture is predominantly water. Spring growth may be only 5% to 10% dry matter. The remainder of the bulk weight is water. A cow consuming 30 pounds of dry matter per day when the moisture content is 90% is also ingesting 270 pounds or about 33 gallons of water, which may meet or exceed her requirement at that time of year. Even after forage matures and dries out to 30% dry matter, she will still be taking in nearly 9 gallons of water.

The impact of moisture content of pasture was really brought home to me during a four-year study where we had installed water meters in 14 stock tanks at FSRC and read the meters every day. When pastures were lush, daily water consumption from the tank was often less than five gallons per day by 1200-pound, lactating cows. On days it rained, there was frequently no water drunk from the tanks, illustrating that cattle can sip a lot of water off the ground. Even heavy dews lowered daily water intake from the tanks.

Many ranchers in the Northern Plains and Canada rely on snow to provide necessary water, while just as many others dispute the reliability of this practice. A long-term project conducted in Alberta has demonstrated the value and adequacy of snow as a water source for dry, pregnant beef cows. I think there is a big difference in the value of snow depending on whether the cattle are grazing stockpiled forage or eating hay.

Cattle that are grazing snow-covered stockpile ingest snow with every bite they take and that is why they can get by without supplemental water. If they are eating hay, they have to literally eat snow by itself to gain the water. This is where failure of snow as a reliable source is likely to occur.

My own experience with snow as a water source came while grazing stockpiled pasture at a location where I had to manually fill tanks from a hydrant each day because of the absence of a winter waterer.

I filled the tank in the morning and went back in the evening expecting to need to refill it. In the evening, the tank was still full but frozen over. I figured the tank froze before the cows had a chance to drink so I broke the ice and called the cows. They showed no interest in drinking.

After doing this for a few days, I decided they were getting enough from the snow and quit worrying about them. Since that time I have not worried about providing water to cows grazing snow-covered stockpile. When I have tried to do the same for cows eating hay with snow all around, I found them mobbing the tank when I came to fill it.

I knew sheep had lower water requirements than cattle so we frequently ran sheep on winter pasture without providing water as long as snow was available. Even without snow, the dry ewes often came to water only every two or three days. It is critical to watch the animals and monitor their condition when relying on natural water sources in the winter.

A few years ago there was a discussion on one of the livestock bulletin boards that permeate the Internet regarding snow as a water source. One of the big bones of contention was when someone differentiated between wet and dry snow and claimed that wet snow was a better water source than dry snow.

As you might expect, the physics purists lambasted him by pointing out that snow is just frozen water and that snow couldn't be wetter or dryer. A pound of snow contains a pound of water and there is no way to change that. But livestock do perform better on wet snow. Not because it contains more water but because it is physically easier to ingest.

Wet snow clings to stockpiled grass while dry snow filters off when animals take a bite. If your snow is a fine skiing powder, you may need to provide supplemental water. If it is snowball snow that makes projectiles you can hurl intact for fifty yards and stick to the side of a passing pickup truck, then you have good livestock snow.

Surface puddles and dew on the grass provide a surprising amount of water to livestock when they are present. When it's not present, it's not a source. Puddles and dew aren't nearly as fun or interesting as snow so that's about all I have to say on that subject.

Wells and springs are probably the most preferred water sources for livestock because they can provide the cleanest water and are easily developed. Not only is clean water important to your livestock, but it also minimizes potential problems in the water system. Pumps last longer, pipes don't get sediment, and valves work more reliably when water is clean.

When working with wells and springs it is important to know the recharge rate. Recharge rate is generally expressed as gallons per minute or gallons per hour. The number of livestock that can be watered is controlled by two factors: reservoir capacity and recharge rate.

If livestock once deplete the reservoir, you are running only on recharge rate. Recharge rate can be determined easily in shallow wells by pumping down to a certain level and measuring how much time is required for the water to rise to the original level. Calculate the volume of the cylinder being refilled, convert cubic feet to gallons, and divide by the elapsed time. Deep wells are much more challenging to measure recharge because of the difficulty in determining before and after water levels.

If you determine the well capacity is less than expected livestock demand, you will need to either reduce animal numbers or develop an alternative water source. While deep wells and springs are largely unaffected by seasonal rainfall patterns, shallow wells and springs can be very sensitive to fluctuations in the water table. For this reason, you should determine recharge rate at the seasonal low. Unfortunately, seasonally low recharge rates generally occur in late summer and coincide with peak water demand by livestock.

Most of my livestock water experience is with ponds as the water source. Watering livestock directly out of ponds is the norm in many extensive grazing programs. This results in low water quality, rapid deterioration of the pond, and loss of quality aquatic habitat. By fencing livestock away from the ponds and using a pump and pipeline distribution system you can overcome these problems.

For quality water and good aquatic habitat, a deep pond with grassy vegetation right to the water's edge is essential. I would much rather have a deep pond with less surface area than a large, shallow pond. Deeper ponds are cooler, cleaner, and more reliable during hot dry times than are shallow ponds. Shallow ponds have much greater evaporative loss than do deep ponds. If constructing a new pond, try to have at least 12 to 15 feet of water at the deepest point with steep sides. Greater depth is better, but may be difficult to achieve without spending a lot more money.

It is best to draw water from the area between three feet below the surface and three feet above the pond floor. This zone provides the cleanest water and stays at near constant temperature year around. With a deep pond it is a lot easier to stay within this zone.

Clear, flowing streams can provide excellent water quality and quantity for livestock and have been a commonly used source for many years. With increasing environmental concerns for water quality and aquatic habitat, stream use by domestic livestock is becoming less acceptable in many parts of the country. More and more states are requiring streams to be fenced out of the grazing area. Streams can still serve as a water source but the watering site must be moved away from the stream channel.

While conventional electric or solar pumps can be used in streams, there are a couple of water-powered pump designs that can be used efficiently with flowing streams. Ram pumps and sling pumps are common types. Plans and specs are available through Extension and NRCS offices.

Whatever water source you choose to use, its reliability is a critical factor. Always base your evaluation on worst case scenarios and have a back up plan ready to implement if your supply fails.

The Basics:

- Fresh pasture is predominantly water.
- It is critical to watch the animals and monitor their condition when relying on natural water sources in the winter.
- There is a big difference in the value of snow depending on whether the cattle are grazing stockpiled forage or eating hay.
- Wells and springs are probably the most preferred water sources for livestock because they can provide the cleanest water and are easily developed.

- If you determine the well capacity is less than expected livestock demand, you will need to either reduce animal numbers or develop an alternative water source.
- By fencing livestock away from the ponds and using a pump and pipeline distribution system you can overcome problems associated with ponds.
- Deeper ponds are cooler, cleaner, and more reliable during hot dry times than are shallow ponds.

Think About Your Farm or Ranch:

- What are your water resources?
- Base your evaluation on worst case scenarios and have a back up plan ready to implement if your supply fails.
- When working with wells or springs, the number of livestock that can be watered is controlled by two factors: reservoir capacity and recharge rate. Have you calculated capacity and recharge rate at the seasonal low? Does this meet your livestock's demands?
- Have you fenced off streams or ponds?

48 What You Need to Know about Water Flow

To install a pipeline distribution system or gravity feed a tank behind a pond, the most important concepts to know are pressure, pipe flow, and friction losses.

This should be simple. Water runs downhill, right? Unless you're pumping it uphill or some restriction prevents it from running downhill. There are always exceptions to the rules.

Once you decide to install a pipeline distribution system or gravity feed a tank behind a pond, understanding how water flows is important. The most important concepts to know are pressure, pipe flow, and friction losses.

Pressure is simply the force that makes water move. It can either be created mechanically with a pump or be the result of head pressure. There are several different types of mechanical pumps available and I will not attempt to describe each type or explain where each is the best suited alternative. Work with local technical experts to determine what pumping system is most appropriate for your situation.

Head pressure is created by vertical drop from the water source to the point of use. Water does flow downhill in response to gravity and gravity is what creates head pressure. A vertical column of water 2.3 feet tall creates a pressure of onepound per square inch, the psi that we are all know. As water flows across the landscape in a pipe, it gains or loses approximately 1 psi for every 2.3 feet fall and rise in elevation, respectively. Head pressure is what controls how effectively you can utilize gravity flow water systems.

You can determine head potential using a surveyor's field level by measuring elevation change from the pond surface to the outlet valve. Now that so many of us have hi-tech gadgetry in our tool boxes, a GPS unit can be used to determine elevation along the water line. A moderately priced hunter-sportsman type GPS unit can give elevation accuracy within a couple of feet if it has been properly set up for that location.

Gravity systems work best if the pipe is laid on a downhill grade over its entire course. If there are rises in the line, you will lose pressure potential. Another problem that can arise when gravity flow systems contain undulations is development of air pockets at the high points which can effectively block flow beyond that point. If your water line has a significant rise in it, install a standpipe with removable top or an airlock at the high point of the rise.

If operating a gravity fed water system from a pond, as the water level in the pond drops, the system loses pressure. Many water valves used in stock tanks require a minimum amount of pressure to function properly. There are very low pressure valves available from some companies. If you expect to be operating with low head pressure, make sure to get a low-pressure flow valve.

There are a couple of key points I would like to make regarding pressure and water flow. Some people have the idea that getting a bigger pump or adding more pressure will make water flow anywhere. That is not correct. In most livestock water systems, pipe size is the most restrictive factor affecting water delivery. As water is pushed through a pipe, it drags along the pipe surface creating friction.

Friction is an energy loss and the energy you are losing is the psi pressure. The greater distance you try to push water, the greater the accumulated friction loss. Smaller pipe diameter produces more friction loss compared to larger pipes because more of the water volume in a small pipe is adjacent to the pipe surface, thus a higher percentage of the water flow is being affected. It doesn't matter how big a pump you put on a half-inch pipe, you can't get water through a half-mile of it because of resistance due to friction.

If you increase pressure and try to push the water harder, friction loss also increases. Slow flow has the least friction while rapid flow has the most. Just think about the erosion potential of fast flowing surface streams versus slow flowing streams. It is exactly the same principle, greater transfer of energy with rapid flow produces more erosion.

The most effective way to increase flow rate in a pipeline system is to increase pipe diameter. A big pipe with low pressure will deliver a lot more water at the outlet than will a small pipe with high pressure. I cannot begin to count the times someone has asked me whether I thought they should install a 1 or 1.5 horsepower pump on their water system. After reviewing what they have planned, my usual answer is put in a bigger pipe and use a 3/4-horse pump.

Increasing pipe size from 1 inch to 1.5 inch while maintaining 30 psi pressure produces almost triple the water flow. By comparison, keeping the 1 inch pipe but increasing operating pressure from 30 to 40 psi increases water flow by less than 30%. In a 1 inch pipe, friction loss is nearly doubled going from 30 psi to 40 psi. Once pipe diameter exceeds 1.5 inches, friction loss is negligible at common working pressure ranges. The bottom line is using a bigger pipe solves most water delivery problems.

Also remember that every time you put a sharp bend in the waterline, resistance increases and you lose pressure. Every fitting that is smaller diameter than the pipe is a restriction that increases friction loss. Rolled polyethylene pipe is a popular carrier in livestock water systems. Because the fittings all are inserted into the pipe, PE pipe has much greater friction loss per 1000 feet than does PVC pipe where all the fittings fit on the outside of the pipe. Bottom line is at any comparable pipe size designation, PVC delivers more water than PE pipe.

A lot of people use a much smaller diameter riser coming up from the pipeline to the stock tank. This immediately reduces the amount of water delivered to the tank. If flow rate at the tank is a concern, use as large a diameter riser as feasible. It makes no sense to run 1.5 inch pipe and then use a 3/4 inch riser.

By the same token, using a highly restrictive float valve negates the value of the large pipe installed for the pipeline and riser. Large diameter, full-flow valves should be installed anywhere water demand is high. Yes, they cost more initially, but they save a lot of headaches and retooling later on.

There is a lot of technical support out there available through government agencies, so don't feel like you have to know everything about designing water systems for yourself. It just makes it a lot easier when you understand the basics.

The Basics:

- Pressure is simply the force that makes water move.
- Head pressure is what controls how effectively you can utilize gravity flow water systems.
- Gravity systems work best if the pipe is laid on a downhill grade over its entire course.
- If you expect to be operating with low head pressure, make sure to get a low-pressure flow valve.
- In most livestock water systems, pipe size is the most restrictive factor affecting water delivery.
- The greater distance you try to push water, the greater the accumulated friction loss.
- Slow flow has the least friction while rapid flow has the most.
- Using a bigger pipe solves most water delivery problems.
- At any comparable pipe size designation, PVC delivers more water than PE pipe.
- If flow rate at the tank is a concern, use as large a diameter

riser as feasible.

- Large diameter, full-flow valves should be installed anywhere water demand is high. Yes, they cost more initially, but they save a lot of headaches and retooling later on.

Think About Your Farm or Ranch:

- Determine head pressure by measuring elevation change from pond surface to outlet valve.
- If your water line has a significant rise in it, install a standpipe with removable top or an airlock at the high point of the rise.
- To increase flow rate in a pipeline system, will you need to increase pipe diameter?

49 Positioning Water Tanks and Access Points

Properly constructed permanent watering points offer an easily accessed and reliable water source, as well as physical durability to minimize maintenance.

When you create a permanent watering point, you want it to be a choice you can be satisfied with for quite some time. Location, cost, material, and seasonal suitability all come to mind quickly as choices you need to consider. Do some research and forward planning to come up with the right system for your operation.

While most of our earlier discussion has revolved around pipeline systems and tanks, there is still a time and place for direct livestock access to the water source, whether it be a stream or reservoir.

When I consider direct access, I generally also think of limited or controlled access. It is the continuous presence of animals on the entire stream channel or pond bank that leads to deterioration of the water source.

If an environmentally sensitive stream is the only source of livestock water, the first choice would be to draw water from the stream to a tank located away from the channel area. A pretty good second choice is to develop controlled access drinking points. These are generally constructed from a geo-textile material overlain with rock or other aggregate. The entire stream area is fenced off except for the access area, which forms a salient point into the stream channel. If properly constructed, animal impact on stream quality is very minimal.

Plans for building controlled access drinking points are available from your local NRCS office. One of the very best plans was developed by the folks at Samuel Noble Foundation in Ardmore, OK. This system features a floating PVC pipe frame with electric fencing which automatically adjusts to changing water level.

The environmental advantage to controlled access points is that as little as 1% of the stream bank or pond perimeter may be all that is exposed to the animals. If the livestock are always within 800 to 1000 feet of water, use the 1 ½ inch per animal guideline for calculating how large the access area should be. At greater travel distances, figure about three inches per animal.

Choose a location for the access point that has gentle slope down to the water. Avoid areas with poor drainage or steep grades. Try to keep away from shady areas that can act as an additional attraction for the animals. If shade is near the water point, manure will accumulate there and stream contamination will be much greater than if the animals are just coming for a drink and then leaving.

One of the great advantages of pipelines and tanks is that you can choose the best locations to situate water. Using natural water sources limits your options for location. The best locations for tanks are on higher, well drained sites well away from shade. There is less chance of mud wallows developing on higher sites. Another advantage is that livestock are always moving their manure higher on the landscape.

Gravity-fed tanks located directly behind pond dams draw nutrient loading directly into the surface water drainage system. Continuously grazed pastures with well developed cattle paths leading to the stock tank are little more than flush systems pouring nutrients, sediment, and fecal coliforms directly into the water system. Getting tanks well away from low areas and drainage ways is a good idea.

Permanent watering points can be very expensive or quite affordable, depending on your choice of materials. Prepoured concrete stock tanks are durable, easily cleaned, but come at a very high price. It is not uncommon to get $400 to $1000 invested per tank, depending on size and pad area.

Steel, plastic, and fiberglass tanks are somewhat lower in cost but do not have the longevity of concrete. The most popular tanks among the graziers I have long associated with are tire tanks. These are stock tanks made from used construction equipment or combine tires.

Disposing of tires is a liability for construction and mining companies and in most areas they are happy to give you the tires. Your cost is usually transportation and installation.

The tires are quite heavy and hauling them is not something you do in a 1/4 ton pickup. Individual tires may weigh from 1000 to 3000 lbs, depending on size. Make sure you are equipped to handle this much weight with your trailer and loader.

Tire tanks are cheap and just about indestructible. Full construction details are available from many NRCS and Extension offices.

For those not familiar with tire tanks at all, picture a six-foot diameter tire about 2 ½ feet deep. Cut the upper bead out back to within five or six inches of the tread. This is not something you do with your pocketknife. For most people the preferred tool is a reciprocating saw with a steel or masonry blade, although some use a chainsaw.

It is best to use a loader or hoist to lift up the section you are cutting to keep the saw from binding. Squirt liberally with dish detergent for lubrication as you cut. Avoid steel-belted tires for ease of cutting.

That cutout created the top side of the tank. The riser pipe and overflow drain need to be plumbed up from the ground like any other tank would be. Set the tire over the plumbing.

The bottom of the tire can be sealed in one of two ways.

The original method and still most common is to pour the bottom wheel opening full of concrete up to the bottom bead. Some people pour the entire lower sidewall full of concrete but that triples the amount of concrete required.

It generally takes about a third of a yard to seal most tires. Use a good soupy mix that will spread nicely into the nooks and crannies on the underside of the tire. If you aren't confident in the cement seal, use a good quality silicone sealant, butyl rubber caulk, or plastic roof cement to seal around the tire edge and pipes. I have heard of very few sealing failures with just the concrete.

The second sealing method was probably invented by someone who had a plumbing failure just underneath the tank. To get at the problem, he likely had to chip all the concrete out of the bottom.

The second method involves a plastic plate bolted to the bottom bead and sealed with a water proof sealant as listed above. The ones I have seen used ½ inch thick recycled-plastic construction panels cut to about two inches greater diameter than the wheel opening. Bolt holes were drilled at three-inch intervals, the bead liberally coated with sealant, and the bolts tightened down. Holes were cut for the pipes to enter beforehand and sealed when the plate was put in place. This method allows removal of the tank bottom to access the plumbing below the tank if the need ever arises.

Regardless of the type of tank or waterer used, having good drainage away from the site is critical. To provide solid footing for the livestock, create a drinking pad around the waterer. A lot of concrete pads used to be put in, but they are quite expensive these days. A very durable pad can be created using geo-textile and aggregate. The minimum pad size should allow mature animals to stand fully on it while drinking. Ideally, at least two animal lengths is better. The pad should not drop off abruptly, rather, it should grade into the surrounding area.

Because tank installations are costly, we would typically like them to serve more than one paddock. We used to do this easily by just splitting the tank with a fence run over the top. That immediately cuts your access to tank perimeter in half, meaning your tanks need to be twice as big as originally calculated for the herd. Quartering a tank is even worse.

A better way to approach it is to create a watering block. With this approach, the tank sits in the center of a fenced area. We usually do this with just single strand electric fence and make the width of the area either five times the diameter of the tank or the tank diameter plus two animal lengths. The basic idea is to allow animals to circle the tank and animals drinking there without having to bump or push them. If the area is made much larger it becomes an attractive lounging area for the whole herd and trampling and manure accumulation become more of a problem.

This system works well for two to four paddocks per tank. It can be used for a few more than four paddocks, but crowding at the paddock-to-water area access can become congested and trampling extends out into the paddock creating more damaged and wasted area.

In mixed-species grazing systems, providing ready access to water by the smaller animals is sometimes challenging. Sheep and goats, especially young ones, have a problem drinking from full size cattle tanks. Using tanks with lower level hog waterers is one option. Banking up soil or aggregate around the tank to lower the outside height is another option but it results in more rapid deterioration of steel tanks and makes cattle more likely to fall in the tank unless a barrier is built over the tank.

Properly constructed permanent watering points offer the animals an easily accessed and reliable water source, as well as physical durability to minimize your maintenance input. Even the best constructed site will need some attention over the years so don't just assume it will always be okay. Maintenance done when it is first needed avoids big messes later on.

The Basics:

- If an environmentally sensitive stream is the only source of livestock water, the first choice would be to draw water from the stream to a tank located away from the channel area. A pretty good second choice is to develop controlled access drinking points.
- Choose a location for the access point that has gentle slope down to the water.
- Avoid areas with poor drainage or steep grades.
- Try to keep away from shady areas that can act as an additional attraction for the animals.
- The best locations for tanks are on higher, well drained sites well away from shade.
- Getting tanks well away from low areas and drainage ways is a good idea.
- Having good drainage away from the site is critical.
- To provide solid footing for the livestock create a drinking pad around the waterer.

Think About Your Farm or Ranch:

- In planning permanent watering sites, consider location, cost, material, and seasonal suitability.
- Do some researching and planning for the right system for your operation.
- Check with your local NRCS office or the Noble Foundation in Ardmore, Oklahoma, for plans for building controlled access drinking points.
- Check out construction plans for tire tanks available from many NRCS and Extension offices.

Designing Your Grazing System

50 Begin by Knowing Your Resources

There is no virtue in having more paddocks than anyone else in the county if it doesn't move your operation toward your goals.

The first thing to remember when you begin to layout your grazing system is fence and water are tools to help you accomplish your goals. The grazing system itself is not the goal. There is no virtue in having more paddocks than anyone else in the county if it doesn't move your operation toward your goals. Success is not measured by how many miles of fence you've strung or how many miles of pipeline you've laid. Success is measured by the sustainability of your operation.

Know your goals and design a system to help you get there. In the very first chapter, we discussed the reason for subdividing pastures. It is to gain management control over your resources. First decide what resources you need to control and, second, how tight a rein you want to hold. Those are the factors that will define the appearance of your grazing system.

The first step in designing the system is making a thorough inventory of your resource base. This includes the natural resources of land, water, plants, and animals; the physical resources of facilities and equipment; the human resources or management, labor, and technical support; and capital resources. Maps are good tools for recording information and building inventories. For the computer buffs, there are several mapping software packages available, though none are cheap. Aerial photos available from NRCS are easy to use. At the far end of the spectrum is a blank piece of paper and a pencil.

Different pieces of land have unique characteristics that create opportunities and challenges in management. Understanding the unique characteristics of the land you are operating allows you to fine tune your management. Soil surveys, soil tests, and just getting out there and walking on the land to learn its nature are important steps in coming to know and understand your land. Soil surveys can provide information on the physical characteristics of soils such as infiltration rate and water holding capacity.

Soil types have a lot of bearing on what plants will grow there the best and what times of year may be optimal for grazing those areas. The soil map can help you identify the primary subdivisions for your farm based on soil type and topography.

Soil testing is a valuable tool for identifying both high and low fertility opportunities and challenges. Soil testing should be done systematically based on planned management units. Topography, drainage, and past use are the main determinants of soil fertility variations so consider those factors when planning your soil sampling. Because of the inherent variability in nutrient distribution in pastures, each soil sample should consist of at least 20 soil cores evenly distributed over the sampling area. Most samples should represent no more than 10 to 20 acres each. The soil test results you get back are only as good as the sample you send in.

Identify all potential water sources and their reliability. Wells, springs, streams, lakes and ponds, and public water supplies are possible water sources for livestock. With increasing environmental concerns, your impact on the water supply is as much an issue as the availability for livestock. Before beginning any water development project, make sure you understand the potential environmental impact of what you are doing. A water source should be evaluated based on its reliability in worst case scenario, not average or best-case.

Determine what plants are present in your pastures and their suitability for your goals. Most people can't identify more than a half dozen of the most common forage species, so get a good plant identification guide. Guides are usually available through either the county Extension office or the NRCS. You don't need to know every plant in the pasture, but you should be able to identify the more commonly occurring species. Remember, you can't manage it if you don't know what it is.

It is fairly easy to inventory your livestock, but I think it is also important to inventory wildlife resources for two reasons. First, it shows the public you have a concern for the environment and it allows you to document that your management practices are wildlife friendly. Second, wildlife are a marketable product of the farm or ranch. Once again it isn't necessary to be able to identify every bird and insect, but having a recognition of what is present on the farm and where it is commonly found is a good idea.

Physical facilities include existing fences, watering points, handling facilities, and buildings. Note the location and condition of existing facilities. In the eastern quarter of the USA, an original survey was by mete and bounds - the old English system using landmarks and landscape features to identify property bounds. West of the Appalachian foothills, most of the surveys were changed to the rectangular coordinate system where land is typically divided in increments of square 40-acre parcels. One square mile consists of sixteen 40-acre squares.

While the rectangular system makes figuring land areas very simple, most existing fences are placed arbitrarily on "40 lines" and have little relevance to managing the landscape optimally. If fences are of questionable integrity at all, I advocate getting rid of all of those old paradigm fences and start working from scratch on dividing the landscape into meaningful management units.

Check the reliability of watering facilities and determine

their suitability for your operation. If you want to raise sheep, tall-sided cattle tanks may have minimal value to you without making alterations. Check flow rate of hydrant and tank valves. What may have worked under previous stocking rate and management, may not be adequate for the system you're planning. Flow rate can be easily determined by timing how long it takes to fill a 5-gallon bucket. The importance of flow rate will become apparent when calculating water demand and designing the water system for the grazing cell.

One of the really nice things about cattle exposed to MiG systems is they tend to be calmer and easier to handle compared to cattle on extensive management systems. This means handling facilities don't need to be quite as heavy duty. Good handling facilities are still essential so evaluate what is there and how it will meet your needs.

While older buildings that have been well maintained may have useful function in the operation, many of the buildings on farms and ranches are too far gone to recover without excessive spending. Make a realistic determination of the buildings' usefulness and value to your operation and decide what needs to be done with them. A barnyard full of deteriorating buildings lowers real estate value and can be hazardous to livestock and people. If you have buildings of particular historical or sentimental value, look for outside interests who may be willing to pay for repair and maintenance.

What equipment is currently in your possession and how necessary is it to the operation? In a true grass-based production system, the need to own any farming equipment is really minimal. Take a good hard look at what you have and if it were sold, how might you better invest that money. A number of Midwest operations that I have worked with financed much of their pasture establishment, fence and water development, or livestock purchases through selling off their farming equipment.

In most of the country, you can get the job done by hiring someone to do the work or renting the equipment less expensively than you can do it by owning the equipment yourself. It takes a big operation to justify owning most of the stuff we all think we need.

Evaluating natural and physical resources is the easy part. Looking at yourself, your family, and your employees in an objective way is a lot tougher. If you are going to be the manager, do you have the necessary skills to accomplish what you need to do? If you answer yes, you had better be right. If you answer no, do you know how to acquire those skills or where to find the person who can do it for you.

If you are also labor as well as management, can you physically get it all done? If not, who is going to do it and what will their compensation be? If you are planning on relying on family labor, are all the family members aware of and committed to the same goals as you? What is the long term ramification of using your kids as slave labor? Will they ever want to come home? Will they want to be a part of the operation in the future? These are the issues that will determine the real success of your operation, not the soil type or water availability.

Lastly, there is the capital resource. How does everything get paid for? Serious, long term financial planning is essential to survival and success in farming or ranching. Tragically, financial planning is what many farmers and ranchers leave for the last. We love the land and the work we do on it and so we put our time and devotion into day-to-day operations. Many of us fear financial planning because it all looks so gloomy, but the point is if you do the planning up-front, the picture need not be gloomy.

Wow! That is a lot of stuff to think about. That's why we call it Management-intensive Grazing.

The Basics:

- Soil testing should be done systematically based on planned management units.
- You can't manage it if you don't know what kind of forages

you have.

- Cattle exposed to MiG systems tend to be calmer and easier to handle compared to cattle on extensive management systems.
- In a true grass-based production system, the need to own any farming equipment is really minimal.
- Family and lifestyle issues will determine the real success of your operation, not the soil type or water availability.

Think About Your Farm or Ranch:

- What are the goals for your operation?
- Inventory your resources.
- Evaluate your water source based on its reliability in a worst case scenario.
- Determine what plants are present in your pastures and their suitability for your goals.
- Inventory wildlife along with livestock.
- Note the location and condition of existing facilities.
- Get rid of all of those old paradigm fences and start working from scratch on dividing the landscape into meaningful management units.
- Evaluate existing handling facilities and determine how they will meet your needs.
- Take a good hard look at your equipment and calculate if it were sold, how you might better invest that money.
- If you are going to be the manager, do you have the necessary skills to accomplish what you need to do? If you answer no, do you know how to acquire those skills or where to find the person who can do it for you.
- If you are also labor as well as management, can you physically get it all done? If not, who is going to do it and what will their compensation be?
- Sit down with your family to do some serious, long term financial planning, essential to survival and success in farming or ranching.

51 Basic Guidelines for an Effective Grazing System

The greater the number of paddocks, the more fine tuned the proportion of grazed acres to hayed acres can become.

Because the grazing system is a reflection of the unique goals and resources of individual farms and ranches, everybody's grazing system looks a little different. There are a few basic guidelines for grazing system designs that apply across a wide range of conditions. These are not "rules" that must be followed, just guidelines that can help the system work more efficiently and effectively.

The first guideline is to keep animals within a reasonable distance of water access. Grazing animals like to drink on a regular basis and to do this they must stay within a certain traveling distance of the water source. As their travel distance to water becomes greater, when they do come to drink, they spend much more lounging time close to water. This results in overgrazing near water as well as increased manure deposition close to water.

In large pastures, we have found that beef cows tend to graze tame pastures more severely within a couple hundred feet of the water source (*see figure*). They tend to graze fairly uniformly from about 200 to 800 feet from water and then leave areas of the pasture beyond 800 feet or so underutilized. I like

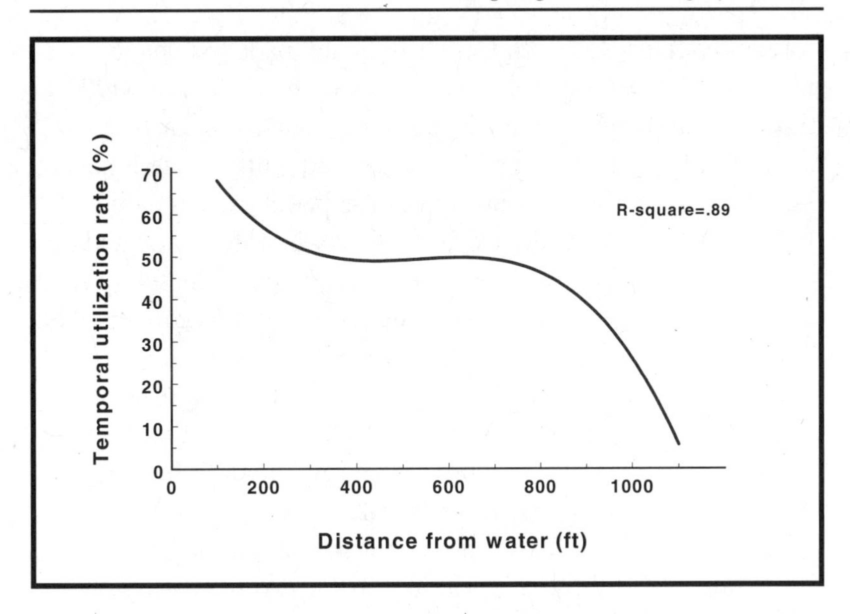

to see watering points in high rainfall or irrigated pastures no more than one-half mile apart, but prefer they be spaced at roughly quarter-mile intervals. If spaced at quarter-mile intervals, grazing animals are much more likely to come to the watering point as individuals or small groups rather than the entire herd.

In rangeland the distance animals will travel comfortably to water is greater due to lesser forage availability. Work done in Texas showed grazing utilization declined when animals traveled over a half mile to water, but the drop in utilization was not economically significant until distance exceeded one mile. Water spacing in rangeland should be two miles or less.

The second guideline is to make paddocks as near to square as possible. This does not mean you should try to make every paddock with four equal-length sides and four right angles to create perfect squares. What it means is avoid long skinny paddocks. Where land allows, uniform sized paddocks with parallel sides are most desirable to allow additional subdivision fencing and mechanical harvest when needed.

Observations from around the world as well as the

Forage Systems Research Center indicate paddock shape influences grazing behavior. Paddocks with low length:width ratios tend to be grazed more uniformly than long, narrow paddocks. Long narrow paddocks are frequently grazed much more heavily in the front portion of the paddock compared to the back part of the paddock, particularly if water is located on one end. This is a common problem on grazing cells set up as a wagon wheel. I recommend keeping length:width ratio less than 4 to 1.

As travel distance from water increases, length:width ratio becomes increasingly important. On small grazing units where livestock are never more than a few hundred feet from the water source, shape is less critical.

Shape is also less critical with shorter grazing periods. Cattle take about three days to establish a strong grazing pattern within an individual paddock. As cattle are allowed to remain on a paddock beyond three days, spot grazing and pronounced cattle trails will begin to develop. When the cattle return to this paddock in future grazing cycles, the grazing pattern is already established and they will begin to follow those patterns on the first and second day of grazing.

The amount of fencing required to subdivide a large tract into paddocks varies depending on the shape and layout of paddocks. Frequently, fencing needs can be reduced by 10-40 percent by considering different layout options.

For any given acreage, a square paddock requires less fence than any other shape, except for a circle. On many farms with rough terrain and dissected by timber and creeks, square paddocks are impossible and numerous trial and error measurements are the only way to find the best design. Using an aerial photo with a clear overlay sheet allow you to look at many options before you ever build the first fence.

That brings us to the third guideline which is to work with the topography to make your primary subdivisions. One of the main goals of grazing management is to utilize the forage resource as efficiently as possible. Because plant community and productivity are closely tied to soil and water conditions, our grazing management needs to be tied to the landscape. The best place to start is to base subdivision on the landscape.

If bottomland and sideslope are included in the same paddock, we frequently see overgrazing on the slopes because pasture regrows more slowly on the slopes. The manager has to make a decision to leave or move the livestock by looking at the paddock. Part is being over-utilized while part is being under-utilized, so what do you do? The best solution is to have the two areas as separate management units.

If a large pasture was seeded down with two distinctively different forage mixtures, the area should be subdivided based on pasture composition. Otherwise, the more preferred mixture will be over-utilized. Avoiding as many of these challenges as possible through good forward planning makes overall management a lot simpler.

Keeping management simple is part of the basis for the fourth guideline, which is to make paddocks of similar grazing capacity, not similar size. More productive areas support higher carrying capacity both on a daily and on an annual basis. If paddocks are all made the same size, more productive paddocks need a longer grazing period while less productive paddocks need fewer days of grazing. To keep the animal's diet consistent, grazing days per paddock should be kept as similar as possible.

Changes in forage availability and quality and animal intake occur at different points in the grazing cycle. Variance in diet from day to day can result in inconsistent animal performance. The nearer to constant the diet can be kept, the more predictable animal performance will be. One of the significant advantages of continuous grazing is that the diet of the animals is kept very constant. Using paddocks of all the same size in acres but that supply a range of two to seven grazing days causes nutritional stress in the animal as intake and quality change through the longer grazing periods.

For these reasons, paddocks on the most productive

part of the grazing unit should be smaller than average while paddocks on the least productive soils will be larger than average. This is why it is important to plan each subdivision fence with stocking rate and animal production in mind, not simply to divide the farm into equal sized units. The more diverse the soil types and terrain on a particular farm, the more useful flexible fence and water systems become.

The fifth guideline is one of my pet peeves. That is to use lanes for animal movement only. Lanes frequently receive criticism as sources of erosion and wasted land. My experience with over 20 years of lane use is that most erosion develops from vehicle traffic in the lane, not animal usage.

If water is available in each paddock, the function of the lane is only to facilitate movement of livestock from one paddock to another. This type of animal usage places much less stress on the vegetative cover in the lane than does daily travel to water or regularly driving trucks or tractors in the lane. We put gates at other points in the fence line to allow vehicle traffic within the grazing cell without driving in the lane.

There really isn't a guideline for number of paddocks in the grazing system. In the "Introduction to Management-intensive Grazing" section we discussed the basis of pasture subdivision. The optimum number of paddocks depends on both pasture species and animal type based on desired utilization and performance goals, resistance to grazing, regrowth habit, and economic potential. For me, the ideal system is moving daily to a fresh paddock. The advantages of daily rotation include minimal feed wastage, consistent forage quality each day, rapid uniform grazing, more uniform manure distribution, and many other benefits that have been presented throughout this book. The main disadvantage is additional cost incurred by the greater level of subdivision.

To even begin to realize many of the advantages of MiG, grazing periods need to be less than four days. To achieve a four-day grazing period during the growing season on tame pasture, a system needs to have a minimum of eight to twelve paddocks to allow the appropriate rest period. If your goal is year-around grazing, many more paddocks will be needed.

The actual number of paddocks required for a particular grazing cycle is determined by the necessary rest interval required for that particular pasture mix under current environmental conditions in combination with the maximum number of days that animals should be left on a paddock to ensure achievement of performance targets. Typically the CHO replenishment cycle in forage plants takes 20-40 days, therefore, this is the range in rest interval we should be generally considering.

Under good growing conditions, a 20-day rest may be plenty whereas in midsummer a cool-season forage may require 40+ days to reach a state of positive CHO balance due to slower regrowth rate. The implication is fewer paddocks or more livestock are needed at certain times of the year. Varying stocking rate to utilize excess spring forage is usually the more cost effective way of dealing with surplus growth. Paddocks not needed for grazing can be harvested as hay or haylage. The greater the number of paddocks, the more fine tuned the proportion of grazed acres to hayed acres can become.

Forward planning is essential to designing a grazing system that will effectively help you accomplish your goals. Doing a good job of assessing your resources and knowing your goals is a good first step toward designing the system. Following these basic guidelines can make the system work more efficiently and make it a little simpler to manage.

The Basics:

- Keep animals within a reasonable distance of water access.
- Make paddocks as near to square as possible. Avoid long skinny paddocks. Paddock shape is also critical with shorter grazing periods.
- Work with the topography to make your primary subdivisions.

- Make paddocks of similar grazing capacity, not similar size.
- Use lanes for animal movement only.

Think About Your Farm or Ranch:

- Plan your subdivisions based on your landscape.
- Plan each subdivision fence with stocking rate and animal production in mind. Don't simply divide the farm into equal sized units.
- Evaluate your grazing goals to determine the number of paddock subdivisions.

52 Advantages in Fixed and Flexible Systems

Your system needs to be a reflection of your goals.

When it comes to making fence and water developments, we have two basic approaches to setting up a grazing system. I like to refer to them as fixed or flexible systems. A fixed system is built using permanent fence and water installations to create the grazing cell while a flexible system relies on movable fence and water for paddock subdivisions within a framework of permanent fence and water installation. Then, there is a wide range of combinations of the two.

Deciding which approach is right for your operation depends on a number of factors. Flexible systems allow much more fine tuning of the pasture-animal balance so if you need tight management control, a flexible system might be up your alley. Fixed systems require less daily labor so, if you're short on time, a fixed system might suit you better.

Any kind of fence or water development costs money. Fixed systems are better suited for large operations where paddocks are large enough that development costs can be spread over many acres.

If you're short on capital, starting out with a flexible

system might be more affordable. There are advantages and disadvantages to each approach.

In a fixed system, all subdivisions are made with permanent fence. My definition of a permanent fence for any class of cattle is a single strand of electrified 12.5 ga hi-tensile wire on solid corners with line posts the deer don't knock the wire off. In very dry conditions, a two-wire fence may be necessary with one strand hot and one strand grounded. For sheep or goats, three-strand fences are most commonly used, although some people do all right with just two wires.

Watering points in fixed systems are commonly either the water source itself, such as spring development or pond, or has water delivered to a tank via buried pipeline. Tanks are typically large, often containing a full day's water supply or more. Pipe flow can be pretty slow and the water supply will still be adequate. The negative effect is watering activities always occur at the same locations in fixed systems.

Fixed systems have a number of advantages, particularly on large operations. Because the expensive part of fencing is in energizers, corners and gates, the more acres you can spread those costs over, the lower the cost per acre. Ultimately it is animal product sold per acre that pays the bills, so lower costs per acre yields higher gross margin per acre.

Picture this. It takes the same number of corners and gates to create ten paddocks on 20 acres as it does to make ten paddocks on 200 acres. Cost per acre is going to be much lower on the 200 acres compared to the 20.

On most days, the labor requirement on a fixed system consists of opening the gate and giving the stock access to the next paddock. Use good quality material for the fence and water systems in the first place, install it properly, and maintenance on the grazing system is pretty minimal. The initial investment may be higher up front, but operating costs on a fixed system can be quite low.

The main disadvantage to fixed systems is that management flexibility is limited. As growth rate changes through the season, you can adjust the length of the grazing period but you cannot change the size of the pasture allocation without resorting to temporary fencing. If paddocks are less than ten acres, efficiency of any mechanical operation is also diminished. On smaller operations, the cost per acre can be substantial because of the limited number of acres over which to spread costs.

Flexible systems require more management and more daily labor to use effectively, but offer several important advantages. The primary advantage is increased management flexibility from both grazing and mechanical harvest perspective. The ability to flex paddock size and forage allocation as conditions change offers greater management control.

Labor requirements in a well designed flexible system can be fairly minimal. Getting an easy working system requires setting up the permanent framework to enhance your time efficiency. Laying out the grazing system as a series of near-parallel grazing corridors is the starting point. I generally recommend keeping corridor width to less than 650 feet to accommodate easy use of fence reels and step-in posts.

My personal preference is to space permanent fences at 435 feet with a water line along alternating fence lines. Why 435 feet? Because each 100 ft allocation equals one acre. It makes it very easy to keep track of what you are allocating and give instructions to kids and hired help doing chores for you. To make it even easier, set line posts in the permanent fences at measured intervals to give known allocation areas.

Even though the daily labor requirement for operating a flexible system is greater than labor in a fixed system, the time needed to move fence and water is pretty minimal if you are using the right equipment and have set the system up as described above.

Over the last 20 years I have timed myself and other people working with portable fence and water systems. Day in and day out, it takes me less than ten minutes to take down a 400 to 500 foot section of polytape on step-in posts and reset it for the next grazing strip. Using a 50 gal plastic tank and quick coupler valves requires about five minutes to move. In most situations it takes longer to get from the house to the paddock than to make the paddock shift.

During one five-year research project at FSRC, we had eight herds rotating in flexible systems. I could typically go out and move all eight herds in about two hours including walking from pasture to pasture. The lengths of fence ranged from 330 to 600 feet and the water tanks were 25 gallon plastic tubs. Set it up right, use the right equipment, and it doesn't take much time at all.

A water line along each alternating fence allows two corridors to be watered from each line. The spacing of the water outlets depends on expected herd size and the needed allocation area. The key factor to making portable water tanks work is high recharge rate in the delivery system.

Whereas water tanks in fixed systems can be large and rely on slower refill rates, flexible systems must be installed with high flow capacity to get rapid refill so that stock can never drain the tank. Some people think this means high pump pressure but it is actually much more related to pipe size. Just increasing pipe size from 1" to 1.25" increases water flow by more than 50%.

Even if you install appropriate sized pipe for the main line, water flow can be restricted by using a small diameter delivery hose or a tank valve that restricts water flow excessively. Hydrants and quick coupler valves are the first point of restriction from the pipeline. Use larger diameter hydrants and valves wherever possible.

If you put in a one-inch hydrant, don't lose that advantage by running a half-inch hose to the tank, particularly if using hose lengths greater the 20 to 30 feet. Larger diameter hoses cost more but deliver a lot more water. A one inch hose delivers roughly four times as much water as a half-inch hose.

So how do you decide whether to use a fixed or flexible approach?

As a general guideline, I suggest flexible if your operation is less than 160 acres. If you are managing over 1000 acres in a single grazing system, fixed is most likely your best option. In between those two benchmarks, a combination system is likely to work best. Having said that, I know operations in excess of 1000 acres that are almost entirely flexible and I know 40 acre farms set up with fixed facilities.

Dennis McDonald, an early pioneer of controlled grazing in north Missouri, originally set up all of his farms as fixed systems. Dennis runs several hundred head of cattle on over a thousand acres. After 10 to 12 years of grazing successfully in those fixed systems, he converted all of his farms to flexible systems because he saw even more benefit to tighter control and found the flexible system easier to manage. Dennis is a full-time farmer who markets his labor through cattle production. The additional time he spends managing in the flexible mode has increased his output per acre so it is time well spent.

Jim Twesten is a custom stocker grazier in north Missouri who works off the farm and has very limited daylight hours on his home place. Jim typically runs around 300 head of yearlings on his 320-acre home unit in a fixed system consisting of about 20 paddocks with permanent water tanks. Because time is his most precious commodity, Jim prefers to minimize his daily labor at home and uses a well designed fixed system.

Eric Bright operates a 60-cow pasture-based dairy in our area and runs a very intensive flexible system to maximize grazing intake of his dairy cows. Moving fence and water every 12 hours requires using the right tools to do it efficiently. Eric uses some annual crops such as ryegrass and brown midrib sorghum so the flexible setup allows him to no-till drill the annual crops in long corridors rather than in small blocks.

Three graziers with three different sets of goals and you get three different systems. Your system needs to be a reflection of your goals.

The Basics:

- Flexible systems allow much more fine tuning of the pasture-animal balance.
- Fixed systems require less daily labor.
- Fixed systems are better suited for large operations where paddocks are large enough that development costs can be spread over many acres.
- Starting out, a flexible system might be more affordable.
- Animal product sold per acre pays the bills, so lower cost per acre yields higher gross margin per acre.
- The main disadvantage to fixed systems is that management flexibility is limited.
- The ability to flex paddock size and forage allocation as conditions change offers greater management control in flexible systems.
- Set line posts in the permanent fences at measured intervals to give known allocation areas.
- Set it up right, use the right equipment, and it doesn't take much time at all.

Think About Your Farm or Ranch:

- Will a fixed or flexible grazing system work best for you?
- Once again, what are your goals?

Putting It All Together

53 Use Pasture Records for Information

Four critical pieces of information that should be recorded in a grazing operation are available forage, pasture condition, animal output, and pasture costs.

Keeping records is not something many of us enjoy. Calving records and herd performance records aren't so bad, but grass records? You have got to be kidding!

Whether or not you want to keep pasture records depends on whether you are just generating data or actually creating information. There is a big difference. You can collect forage production and carrying capacity records for every paddock, accumulate it all in a three ring binder or in a spreadsheet, but until you use it to help make a business decision, you just have data. Unless it becomes useful information for your business, there is no point in keeping records. Once you make the decision to use the data you collect, then you need to decide what data is important.

To make useful business decisions, I think there are four critical pieces of information that should be recorded in a grazing operation: available forage, pasture condition, animal output, and pasture costs. These are fairly broad terms that can be narrowed down to fit your particular goals and needs. Available forage is a measurement of how much grass is available for grazing on the entire farm or in a particular paddock. In some parts of the world, pasture cover is the commonly used term.

In chapter 22 we discussed a number of different ways of estimating available forage. Any of them will work, it just depends on how much you're willing to spend. My preference is predicting forage yield from sward height. Properly done it is as accurate as more time consuming methods with minimal cost involved. There are a couple of ways to approach available forage recording. One is to record yield on a particular paddock only when stock are turned to that paddock. Combined with measurement of post-grazing residual, this can provide some valuable information, including available forage at turn in, residual, utilization rate, and regrowth rate. What it doesn't do is keep you informed of what your current forage supply is.

The other approach is to measure every paddock bi-weekly or at some other set frequency. This approach provides an ongoing inventory of forage supply which can really help planning marketing and purchasing decisions.

With either of these systems, I like to use a spreadsheet program to keep the data and generate the information I need for evaluating management decisions. Each paddock has an individual record so paddocks which have been overseeded or fertilized, grazed at certain times of the year, or any number of other factors can be compared to other paddocks. Costs for each practice can be compared and cost:benefit ratios determined. We always talk about getting the biggest bang for your buck, this is how you measure it. For non-computer users, traditional accounting sheets can be set up and the same information recorded, it is just not as convenient, which means it is less likely to get done.

Pasture condition includes such things as plant vigor, biodiversity, ground cover, weed infestation, and soil conditions. These are all indicators of pasture health. Monitoring them over time can tell you whether your pastures are getting better or worse or staying about the same.

A really useful sheet for recording and keeping this information is available through the National Resource Conservation Service (NRCS) in most states. Check at your local NRCS office for the Pasture Condition Worksheet. The sheets are pretty self explanatory and can be worked through fairly quickly with just a little experience. I like doing pasture condition score sheets twice a year, once in late spring or early summer and again in the late summer or early autumn. This allows you to check short term effects that occurred during the growing season. By doing it for multiple years, it allows you to monitor response to changes in management and effects of weather.

Animal output ultimately comes down to pounds of meat, milk, or other product sold from the farm or ranch. Almost any avenue you sell these products through provides a receipt of pounds or head marketed. You might want to keep track of animal use of individual paddocks or grazing cells.

Just keeping a record of animal unit days (AUDs) or a similar animal use measure can give a good assessment of the production from different grazing units or seasonal distribution of usage. I tend to use a conceptual "cow-day" equivalent based on weight of cows in my herd. Almost all my forage budgeting and record keeping is based on cow-day equivalents. For many this is easier to grasp than standard animal units. The important information is how many head of what class for how long.

If you pay your bills regularly, you should have a record of pasture costs. Whether it is your checkbook, credit card statement, or your QuickBooks® A9 file, financial records are available. The challenge for most people is assigning costs to the appropriate enterprise. This is where computer accounting packages can make record keeping much easier. Different portions of single checks or sales receipts can be assigned to different enterprises and, with a push of the button, total expenses for a particular enterprise are calculated.

So now that you have it, what can you do with it? The first thing is to look at your cost of production for whatever product you're selling. If it is lower than what you're selling your product for, congratulations! You're doing better than most. But as there is always room for improvement, look for where you're spending most of your money. Think about it long and hard. Is it worth it? Are you getting enough return on your investment to justify the expense? If you don't think so, consider the alternatives.

If you have kept animal use or forage production records on individual paddocks along with costs for each paddock, you can determine which pastures and what management strategies are giving you the lowest cost of production. You can begin to move more of your operation toward those production systems. Did putting on 80 pounds of N fertilizer pay? Did adding 6000 feet of waterline pay? You can only answer these questions if you have information.

The Basics:

- Available forage is a measurement of how much grass is available for grazing on the entire farm or in a particular paddock.
- Pasture condition includes such things as plant vigor, biodiversity, ground cover, weed infestation, and soil conditions—indicators of pasture health.
- Animal output ultimately comes down to pounds of meat, milk, or other product sold from the farm or ranch.
- A record of animal unit days or a similar animal use measure can give a good assessment of the production from different grazing units or seasonal distribution of usage. The important information is how many head of what class for how long.

Think About Your Farm or Ranch:

- You already have financial records, figure out how to use them for business decisions.
- Create a spreadsheet program to keep the data and generate the information you need for evaluating management decisions. Set up each paddock as an individual record.

- Plan to create pasture condition score sheets twice a year.
- Keep a record of pasture costs, assigning appropriate expenses to each enterprise.

54 Year Around Grazing is a Reality

You don't have to reduce your cow herd, you just need to change the pattern of use.

Ten years ago I was not as enthusiastic about year-around grazing north of the Mason-Dixon line as I have become more recently. I can remember a number of different times when Bob Evans chastised me for not being more aggressive about year-around grazing.

While I always thought it was a nice concept, I wasn't really sure it was economically feasible on a consistent basis outside of the southern states. Either we couldn't accumulate enough winter pasture to carry a cow through the entire winter, or snow and ice would finally make grazing non-tenable.

I actually got over the ice and snow challenge several years ago when I discovered that cows that had gotten used to grazing in the winter would much rather graze even buried stockpile pasture than eat hay. Maybe that is a real testimony to the poor quality hay most of us put up.

Then, over the last ten years, we learned so much more about how to consistently grow high quality stockpile and how to utilize it efficiently. I used to question the economics of applying 60 pounds of N in August to grow winter pasture. Now I don't hesitate to apply 80 to 100 pounds per acre, because I know it is still a lot cheaper than feeding hay.

Research conducted by Dr. Rob Kallenbach, state extension forage specialist with the University of Missouri, has shown a near linear yield response up to 100 lb N/acre for stockpiling fescue. That means the 100th lb is giving you nearly the same return as the first lb. That is darn good N efficiency.

I need to clarify that statement a little bit. There are areas in the fescue belt where water holding capacity of the soil limits fall pasture yield to a certain level. Applying more N in those areas won't be beneficial.

In the deep rolling prairie and glade soils through much of the lower Midwest and upper South, the productive capacity of the soil to respond to 80 to 100 lbs of N is there. Stockpiling tall fescue for winter grazing is not a new idea. It has been standard practice at the Forage Systems Research Center for over thirty years.

In those early days of winter grazing research, typical yields were just over one ton/acre and 40 to 50 grazing days / acre is all we expected. To graze all winter at those production levels required three or more acres to carry a cow through the winter. That is where economic efficiency went out the window.

The most common question I was asked 20 years ago when I talked to producers about stockpiling was how much they were going to have to reduce their cow herds to be able to have the required winter acreage.

I never had a good answer for them. Now, I know you don't have to reduce your cow herd at all, you just need to change the pattern of use.

Today our stockpile yields are commonly around two tons/acre and we expect to achieve 100 to 120 grazing days/ acre with dry spring calving cows. With improved growing season grazing management, we don't have to begin grazing stockpiled pasture as early in the fall. Whereas we used to think of the winter season as 150 days because that is about how many days grass doesn't grow in north Missouri; we now think of winter as 120 to 130 days.

In this new scenario, it only takes between 1 and 1 ½ acres to carry a dry cow through the winter. Combine stockpiling with 60 days of stalk grazing in the lower Cornbelt and a cow can be carried through the winter with less than an acre of stockpiled fescue.

With our typical N, temporary fence, and labor costs, we can winter a dry cow for between $30 - $40.

While that is all exciting, the real excitement for me has come with fall-calving cows. For years I believed those folks who said fall calving didn't work up here because it costs too much to feed a lactating cow all winter.

As I write this on March 28th 2003, we have just finished our second year of a fall calving study at FSRC. One-third of our cows started the winter on winter annual pastures, one-third grazed stockpiled tall fescue, and the other third were fed hay all winter with all treatments beginning November 12.

I need to mention that it has been extremely dry all across north Missouri since mid-May, 2002.

Our stockpile was from 40% to 80% of normal depending on when the N was applied. Even with these conditions the cows grazing stockpile have gone the entire winter all the way through to spring grass with no supplemental feed.

On this day of March 28, the calves averaged 502 lbs, the cows are still above body condition score 6, and all at a cost of less than $45 per cow. And here we sit with 5-weight cattle ready to go to lush pasture to make their cheapest gain of the year.

After years of telling other graziers that year-around grazing was a real possibility, but never having done it myself, I can now say that I have taken a herd of cows through an entire winter with pasture only. Coincidentally, March 28, 2003, was also my last day of working at the Forage Systems Research Center. After 22 years of working on improving efficiency of beef operations, March 28 was as fitting an end to my research career as I ever could have asked for.

The Basics:

- Cows can go through an entire winter on grazing alone.

Think About Your Farm or Ranch:

- Are you applying the right amount of N for your pasture conditions?
- What can you do to extend your grazing year-round?

55 Bringing a Dead Farm Back to Life

Figuring out early on how to vary stocking rate to fully utilize the spring flush and avoid making hay is critical to building vigor on a dead farm.

Traveling is one of my favorite pastimes. I never get tired of just looking at the landscape whether it be farms or wilderness. Towns and cities get me down a little, but most land I enjoy seeing.

As you might imagine, looking at grass farms are some of my favorite scenery. One of the more interesting things to me is the variance you can see from one farm to another.

I remember many years ago driving through the southern Flint Hills in Kansas for the first time. I had heard about how great Flint Hills range was for many years. What I saw was largely overgrazed and weedy remnants of what had once been the great tall grass prairie.

Then I came over a hill and saw real grass, a true night and day difference. That high quality, healthy rangeland went on for a number of miles and then the scenery deteriorated again.

I found out a few years later I had been driving through the DeVore Ranch, one of the few Holistic Resource Management practitioners in the Flint Hills at that time.

I had a similar experience driving through Queensland, Australia, as we passed from the typical range country onto stations practicing controlled grazing. And many of you have heard me tell stories about the fenceline differences between our farm and some neighboring properties. The difference between the good grass and the poor grass scenery is the good grassland is a healthy, living, breathing organism while the poor grassland is dead.

Much farmland is dead from over-tillage and erosion. Much pastureland is dead from overgrazing, erosion, and lack of fertility. Both situations suffer from the same basic problem: depleted organic matter.

Part of my education over the last 20 years has come from the fact Dawn and I had bought a dead farm. Of our 260 acres, about 200 acres had been continuously row cropped for many years. The topsoil had been eroded away and much of the landscape showed the yellow and red clay subsoil. Soil pH ranged from 4.3 to 5.5, with a lot more acres below 5 than above 5.

The previous owner had abandoned some areas all together because they could no longer grow any crops worth harvesting. In summary, it was not very pretty.

The one bright spot was we bought most of it in 1986 at the deepest depths of the farm crisis of the '80s and we bought it cheap. But we faced the challenge of how to bring it back into productive grassland. The following is a retrospective look and what we did and didn't do.

I am not sure who actually first coined the phrase, "Plant nothing but fence posts the first year," but it is as true an axiom of farm reclamation as any I know. Livestock can and do graze many weeds quite willingly. Abandoned cropland will usually grow some weeds. While not all of them will be palatable, most are, especially if grazed at very immature stages.

Seeding down land is an expensive proposition. We couldn't afford to do very many acres at a time. It took us 12 years to seed down roughly 200 acres. Some years we didn't seed any new ground and some years we did 40 plus acres. It all depended on cash flow and weather conditions. Borrowing money to seed pasture is way too risky in my book and not many bankers are willing to finance seeding pasture because it is not a very liquid asset.

If you want to borrow money for something, buy cattle to graze weeds. A better option is to take a lesson from Greg Judy, whose book *No Risk Ranching* explains how to contract graze while trying to build your business.

Our farm had very low soil pH. One thing we didn't do soon enough was to start liming. Once we began our liming program, the improvement was dramatic. I believe, in the eastern half of the USA, lime is the most important fertilizer for most grass farms.

Legumes are critical for profitable grazing operations. Few legumes will thrive in acid soils and thriving legumes are what crank out nitrogen and help rebuild soil organic matter. Liming will improve availability of many other soil nutrients.

In our part of the country, phosphorus is the second most limiting nutrient after N. Applying P fertilizer to soils with pH less than 5 is just about a waste of time. Under acid conditions, P fertilizer is immobilized by chemical reactions with other soil compounds. You need to take care of liming needs before you can get serious about other nutrient management needs.

Once again, liming and fertility improvements for us were governed by cash flow. Every improvement has to be paid for by livestock product. Like many farms and ranches, we were understocked in our beginning years. When cattle prices were low, there weren't many dollars left for lime and P. As stocking rate increased and cattle prices got a little better, we eventually found the necessity to buy lime and fertilizer for tax purposes.

Making hay is a drain on the land as well as the pocketbook. Every bale of hay you harvest in one place and feed somewhere else removes nutrients from the source field. Dead soil can't stand having those nutrients taken away.

Dr. Kevin Moore, University of Missouri economist, reported a number of years ago the most profitable scenario for most beef farms was to stock the farm to its full grazing capacity and buy any needed hay. This recommendation is doubly true for life-depleted grasslands.

Taking hay away is a drain, but bringing hay onto your place can be a much needed transfusion. Not only do you gain the nutrients contained in the hay but feeding hay on poor ground is one of the quickest ways to add some organic matter to the soil and get a little thatch on the soil surface.

I made hay way too long. Figuring out early on how to vary stocking rate to fully utilize the spring flush and avoid making hay is critical to building vigor on a dead farm because most of us can't afford to replace the fertilizer hay is taking off.

During the resurrection process, I tend to view purchased hay as a soil supplement first and as livestock feed second. If it is cheap enough, I would not hesitate to buy junk hay, unroll it, and be content to waste half of it. And lastly, as long as you plan to stay on a particular farm, never, ever sell a bale of hay off that farm.

Always remember you don't have to own the critters on your pastures. We were ten years into our farm before we started contract grazing. The fastest way to get animal numbers up and make a grazing impact is contracting additional livestock onto your place.

To be most successful with contract grazing, forget about those pretty steers. Try to get animals with low nutrient requirements but big capacity. My favorite animal for contract grazing is the dry, pregnant fall-calving cow. They can eat weeds all spring and summer and thrive. The more animals on the place, the more cash flow and the greater the likelihood you can afford to build more fence, lay pipelines, lime and fertilize.

So if you've fenced it, grazed the weeds, started working on the lime and nutrient needs, and have gotten the place stocked to its grazing capacity, then you can think about actually seeding it down to better pasture. Don't waste your time seeding improved pasture species until you have brought the soil back to life. It is a waste of time. Been there, done that.

Sometimes I think if I had it all to do over again, I would pay the price to get a good grass farm to begin with and avoid all the work and expense. But on the other hand, we have created a very nice, productive grass farm through a combination of sound management and judicious spending. Resurrecting a dead row crop farm turned out to be a fairly expensive process, but the education was priceless.

The Basics:

- Plant nothing but fence posts the first year.
- Lime is the most essential pasture fertilizerin the Eastern USA.
- Increase phosphorus levels to ensure legume establishment and persistence.
- Legumes are critical for profitable grazing operations.
- Stock the farm to its full grazing capacity and buy any needed hay.
- If you plan to stay on a particular farm, never, ever sell a bale of hay off that farm.
- The fastest way to get animal numbers up and make a grazing impact is contracting additional livestock onto your place.
- Don't waste your time seeding improved pasture species until you have brought the soil back to life.

Think About Your Farm or Ranch:

- Is all or part of your land dead pasture?
- What caused your pasture to die? Overgrazing, erosion, and lack of fertility?
- What can you do now improve soil fertility without adding much costs?

56 Putting It All Together: Getting Started with MiG

By now you have evaluated your resources and set some goals. Now it is time to start implementing your plans and learn by doing. The only question now is where to begin. Let's start by thinking about learning scenarios.

If you're learning to ride a horse for the first time, it's a good idea to start with a well broke, mature trail horse. Try starting with a raw bucking bronc and you can get hurt mighty bad and you might not ever want to get on another horse as long as you live. On the other hand, if you do manage to stay on the bronc, it can be a real learning experience. The bad thing is you might only know how to ride a wild mustang.

Learning MiG is similar. Trying to start out in the spring on fast growing grass is a real challenge. Add to it changing animal requirements and it can be even more challenging. In the spring we frequently have cows that have just calved. They need to give milk for the calf and get bred back within 90 days of calving so they need some nutritional attention.

The biggest failing I see in MiG operations, even among experienced graziers, is staying too long on a paddock and trying to graze it too short. The outcome is animals not achieving adequate intake and the remaining paddocks getting completely out of control. Spring pasture is riding the bronc. Tough to do unless you have a lot of experience or a natural gift.

I believe **the best time for a new grazier to learn the ropes is grazing winter-dormant pasture with dry stock**. In the eastern half of the USA this is likely to be stockpiled tall fescue or some other cool-season mixed pasture. The majority of cows in that part of the country are still spring-calving so we have an ideal combination of dry cows on dormant pasture. In the South, the pasture might be dormant bermuda grass or slow growing winter annual. Dormant native range might be the learning field on the Plains.

It is really difficult to screw up dormant pasture or a dry, pregnant beef cow. It can be done but you have to work at it. So, assuming you have the winter pasture prepared, let's get started.

Water availability must be considered in all seasons, and winter is no exception. Keeping water open and available during freezing weather can be a challenge. One easy thing about winter water, though, is you don't need to worry about livestock going back across the area they have already grazed and regrazing new shoots. That's why it's hard to screw up dormant pasture. **Plan your grazing sequence to begin near a water source and then work grazing strips away from the water source.**

How much area do you actually need to allocate for each strip? That depends on **how much forage is in the pasture, how much of it you want to utilize, how much each animal needs to eat, how many animals you have, and how long you want them to be there**. In Chapter 21 the formula for calculating appropriate stock density is presented. If you know or can estimate reasonably accurately the five factors listed above, you can use the formula to come up with how many acres to allocate in each grazing strip.

I think it is a good idea to go through that exercise several times just to get a good grasp on what factors are important and how they relate to one another. The seat of the pants approach, which is a good education in itself, is to just go out and put up a temporary fence. Just look at the pasture and estimate how much area your herd needs for one day of grazing and set the fence accordingly. Then go back to the house and don't go back until tomorrow.

When you go back tomorrow, look at the cows and the pasture. If the pasture is grubbed down short and the cows are bawling, you guessed wrong. Today you know you need to give them a bigger area. If there is still a lot of grass in the strip and the cows are happy, you guessed wrong the other way. Leave the animals where they are and come back tomorrow. Estimate how much of the grass is still available and make an adjustment in the next strip by allocating less.

Keep repeating this process and by the end of winter you will be familiar with using movable fence and estimating how much pasture your herd eats in a day. Where you want to end up is being able to look at the pasture and the stock and make an allocation without needing to think about actual numbers. It's all about relating grass to livestock and it is a learned art.

When spring rolls around, things are going to shift into high gear. Grass is going to start growing faster than you can imagine and animal needs are going to change. Obviously you need to adapt. You have just jumped from the old trail nag to the mustang. Here are some things to think about.

If you grew up or have been exposed to continuous grazing, any rules of thumb or guidelines you learned about when to begin grazing in the spring no longer apply. Waiting for cool-season grass to get six inches tall before you turn livestock in is a sure recipe for pastures getting out of control. **Stock need to be out early**. If you have ten or more paddocks, as soon as you think there might be enough grass in two or three paddocks to feed the herd one day each, start grazing. On native tall grasses, start just a little bit later and look for adequate feed in about one-third of your paddocks before you begin grazing.

If you run out of grass in just a few days, you guessed wrong and you can go back to feeding hay. Feeding hay is what you were going to be doing anyway, so don't worry about it. If you see a few more days of grass getting ahead of the stock, keep them moving ahead and get ready to ride the wave. As the years go by and you learn more and more, you will begin to learn some other things to do this time of year to keep a tighter rein on your pastures.

In the first grazing cycle, do not try to utilize all of the grass in the paddock. Remember take half, leave half. In the first cycle, you may take no more than a quarter to a third of the available pasture. Don't get bogged down trying to fully utilize every paddock. **Get across all of them fast and come back to them quickly. Make sure you are leaving adequate residual.**

Pretty soon you see paddocks are getting ahead of the cattle and it's beginning to look a lot more like hay than pasture. You have several choices. You can accelerate your rotation, leaving more and more residual in each pasture for stockpiled summer use. This works fine for beef cows, growing dairy heifers, and sheep, but it doesn't work for stock with high nutrient demand like dairy cows or beef stockers.

Where forage quality is an issue, the runaway paddocks need to be mechanically harvested to get new regrowth coming. The more experience you get, the earlier in the season you will see this juncture coming. The earlier you see it, the earlier you can do something about it. The earlier runaway paddocks get harvested, the more likely they are to make good regrowth. **Quick regrowth after harvest provides more summer feed, which means you will be more likely to graze later into the fall and winter.**

The whole season builds on what happens in the first month or so. This is why you do not want to start your MiG program in the spring. Not only is it challenging, it can also be disappointing and disillusioning. Not at all where you want to be.

How can you accelerate your learning process and avoid making some of the more common mistakes? Hanging out with the right kind of people is a great first step. **Next to making all the mistakes yourself, the next best education is pasture walks with other graziers.** Seeing what other grass-minded people are doing and learning from their experiences is not only valuable but can also be a lot of fun.

I have had the privilege of attending pasture walks or their equivalent all across the USA, as well as in Australia, Brazil, Canada, Ireland, and New Zealand. I don't know what it is about graziers, but anywhere in the world I have found they are friendly and open about their experiences. Any chance you have, take advantage of that resource. If you don't have time to go on an occasional pasture walk, a reexamination of your priorities might be in order.

Grazing conferences, seminars, and short courses can also be helpful. Programs that incorporate either field sessions or producer presentations tend to be much more useful than all academic, classroom programs. And that comes from a guy who spent 20-plus years in academia. Nothing replaces real world experience.

It's in your hands now.

Good luck and good grazing.

Grazier's Glossary

AI: Artificial insemination.

Aftermath: Forage that is left or grown after a machine harvest such as corn stalks or volunteer wheat or oats. Also called the "Fat of the Land."

Animal unit day: Amount of forage necessary to graze one animal unit (one dry 1100 lb beef cow) for one day.

Annual leys: Temporary pastures of annual forage crops such as annual ryegrass, oats or sorghum-sudangrass.

AU Animal unit: One mature, non-lactating cow weighing 1100 lbs or its weight and class equivalent in other species. (Example: 10 dry ewes equal one animal unit.)

AUD: Animal unit day. Amount of forage needed to graze one animal unit for one day.

AUM: Animal unit month. Amount of forage needed to graze one animal unit for a month.

BCS: Body condition score. A 9-point scale to describe beef cow condition.

Blaze graze: A very fast rotation used in the spring to prevent the grass from forming a seedhead. Usually used with dairy cattle.

Body condition score: BCS. A 9-point scale to describe beef cow condition.

Break grazing: The apportioning of a small piece of a larger paddock with temporary fence for rationing or utilization purposes.

Breaks: An apportionment of a paddock with temporary electric fence. The moving of the forward wire would create a "fresh break" of grass for the animals.

Carrying capacity: Stocking rate at which animal performance goals can be achieved while maintaining the integrity of the resource base.

Cash fat cattle: These are real cattle selling for real money in real time.

Cattle cycle: The ten to twelve year price cycle from peak to valley for breeding stock.

CDA: Cow-days/acre. The estimated number of cows that could graze the standing forage on one acre for one day.

CHO: Carbohydrates.

Clamp: A temporary polyethylene covered silage stack made in the pasture without permanent sides or structures.

Composting: The mixing of animal manure with a carbon source under a damp, aerobic environment so as to stabilize and enhance the nutrients in the manure.

Continuous grazing: See below.

Continuous stocking: Allowing the animals access to an entire pasture for a long period without paddock rotation.

Coppice: Young regrowth on a cut tree or bush.

Compensatory gain: The rapid weight gain experienced by animals when allowed access to plentiful high quality forage after a period of rationed feed. Animals that are wintered at low rates of gain and are allowed to compensate in the spring frequently weigh almost the same by mid-summer as those managed through the winter at a high rate of gain. Also known as "pop."

Creep grazing: Allowing calves to graze ahead of their mothers by keeping the forward paddock wire high enough for the calves to go under but low enough to restrain the cows.

CRP: Conservation Reserve Program.

CWT: 100 pounds.

Deferred grazing: The dropping of a paddock from a rotation for use at a later time.

Dirty fescue: Fescue containing an endophytic fungus which lowers the animal's ability to deal with heat. Fescue without this endophyte is called Fungus-free or Endophyte-free.

Dry matter: Forage after the moisture has been removed.

EPD: Expected Progeny Difference. A statistical expression of expected genetic effect of a known sire on his progeny.

Fat cattle: This was the original term for cattle that are "finished" and ready for slaughter. To be politically correct in this low-fat world, this term has been changed to Live Cattle by the Chicago Mercantile Exchange.

Fats: The same as fat cattle, live cattle and fed cattle. Indicates cattle are ready for slaughter.

Fed cattle: Used interchangeably with Fat or Live Cattle. This indicates they have been in a feedlot.

Feeder cattle: These are cattle weighing between 700 and 850 pounds.

Flogging: The grazing of a paddock to a very low residual. This is frequently done in the winter to stimulate clover growth the following spring.

Forbs: General term used to describe broad-leafed plants.

Frontal grazing: An Argentine grazing method whereby the animals grazing speed is determined with the use of a grazing speed governor on a sliding fence.

FSRC: Forage Systems Research Center located at the University of Missouri in Linneus.

Grazer: A animal that gathers its food by grazing.

Grazier: A human who manages grazing animals.

Grazing pressure: How deep into the plant canopy the animals will graze.

Heavy feeders: Feeder cattle weighing over 800 pounds.

Heavy weights: In fat cattle, these would weigh over 1250 pounds.

Herd effect: Animal concentration that creates some impact on the landscape or environment.

Intake: Amount of forage an animal will consume. May be expressed as pounds of dry matter/head or as a percent of animal liveweight.

K: Potassium.

LAI: Leaf area index. A term used to describe the relative amount of leaf area-to-ground area.

Lax grazing: The allowing of the animal to have a high degree of selectivity in their grazing. Lax grazing is used when a very high level of animal performance is desired.

Leader:follower: A leader:follower grazing system is one in which two or more classes of livestock having distinctly different nutritional needs or grazing habits are grazed successively in a pasture.

Leader:follower grazing: The use of a high production class of animal followed by a lower production class. For example, lactating dairy cattle followed by replacements. This type of grazing allows both a high level of animal performance and a high level of pasture utilization. Also, called first-last grazing.

Leaf area index: See LAI.

Legumes: Plants that bear fruits such as beans or clover. Most legumes have the capability for symbiotic nitrogen fixation.

Ley pasture: Temporary pasture. Usually of annuals.

Lignin: "Woody," non-fiber components of plant cell walls. It is completely indigestible to animals at any stage of a plant's maturity. Lignin concentration increases as plants mature.

Lodged over: Grass that has grown so tall it has fallen over on itself. Most grasses will self-smother when lodged. A major exception is tall fescue and for this reason it is a prized grass for autumn stockpiling.

Management-intensive Grazing (MiG): The thoughtful use of grazing manipulation to produce a desired agronomic and/or animal result. This may include both rotational and continuous stocking depending upon the season.

Mixed grazing: The use of different animal species grazing either together or in a sequence.

Mob grazing: A mob is a group of animals. This term is used to indicate a high stock density.

N: Nitrogen.

OM: Organic matter in the soil.

P: Phosphorus.

Paddock: A permanently fenced pasture subdivision.

Pastureland: Land used primarily for grazing purposes.

Pop: Compensatory Gain.

Popping paddocks: Paddocks of high quality grass and legumes used to maximize compensatory gain in animals before sale or slaughter.

Pugging: Also called bogging. The breaking of the sod's surface by the animals' hooves in wet weather. Can be used as tool for planting new seeds.

Put and take: The adding and subtracting of animals to maintain a desired grass residual and quality. For example, the movement of beef cows from rangeland to keep a rapidly growing tame stocker or dairy pasture from making a seedhead in the spring and thereby losing its quality.

Range: A pasture of native plant species including mixtures of grasses, forbs, and browse.

Rate of passage: The time it takes for forage to pass through the digestive tract and be eliminated from an animal.

Rational grazing: Andre' Voisin's term for Management-intensive Grazing. Rational meant both a thoughtful approach to grazing and a rationing out of the forage for the animal.

Residue: Dead plant litter lying on the soil surface.

Residual: The desired amount of grass to be left in a paddock after grazing. Generally, the higher the grass residual, the higher the animal's rate of gain and milk production. This can be expressed either sward height or forage mass. Example: We could have a four-inch residual or a 1500lb/acre residual.

Rumen: The fermentation vat of a ruminant's stomachs.

Ruminants: Hooved livestock with multiple stomachs and which chew their cud.

Seasonal grazing: Grazing restricted to one season of the year. For example, the use of high mountain pastures in the summer.

Self feeding: Allowing the animals to eat directly from the silage face by means of a rationing electric wire or sliding headgate.

Set stocking: The same as continuous stocking. Small groups of animals are placed in each paddock and not rotated. Frequently used in the spring with beef and sheep to keep rapidly growing pastures under control.

Solar panel: Green and growing leaves of grasses, forbs and legumes.

Split-turn: The grazing of two separate groups of animals during one grazing season rather than only one group. For example, the selling of one set of winter and spring grazed heavy stocker cattle in the early summer and replacing them with lighter cattle for the summer and fall.

Spring flush or lush: The period of very rapid growth of cool season grasses in the spring.

Standing hay: The deferment of seasonally excess grass for later use. Standing hay is traditionally dead grass. Living hay is the same technique but with green, growing grass.

Steers: Castrated male cattle.

Stock density: The number of animals on a given unit of land at any one time. This is traditionally a short-term measurement. This is very different from stocking rate which is a long term measurement of the whole pasture. For example: 200 steers may have a long-term stocking rate of 200 acres, but may for a half a day all be grazed on a single acre. This acre while being grazed would be said to have a stock density of 200 steers to the acre.

Stocker cattle: Animals being grown on pasture between

weaning and final finish. Stocker cattle weights are traditionally from 300 to 550 lbs.

Stocker cow: A young cow less than five years old.

Stocking rate: A measurement of the long-term carrying capacity of a pasture. See stock density.

Stockpiling: The deferment of pasture for use at a later time. Traditionally this is in the autumn. Also known as "autumn saved pasture" or "foggage."

Strip graze: The use of a frequently moved temporary fence to subdivide a paddock into very small breaks. Most often used to ration grass during winter or droughts.

Swath grazing: The cutting and swathing of small grains into large double-size windrows. These windrows are then rationed out to animals during the winter with temporary electric fence. This method of winter feeding is most-often used in cold, dry winter climates.

Transhumance: The moving of animals to and from seasonal range or pasture. For example, the driving of cattle from winter desert range to high mountain summer range.

Value of gain: The net value of gain after the price rollback of light to heavy cattle has been deducted. To find the net value of gain, the total price of the purchased animal is subtracted from the total price of the sold animal. This price is then divided by the number of cwts. of gain. Profitability is governed by the value of gain rather than the selling price per pound of the cattle.

VDMI: Voluntary dry matter intake. What the animal will consume on its own.

Wintergraze: Grazing in the winter season. This can be on autumn saved pasture or on specially planted winter annuals such as cereal rye and annual ryegrass.

Yearling feeder: Cattle older than weaned calves. A true yearling would be older than one year of age and less than two years of age. Most cattle will weigh between 600 and 750 pounds at one year of age.

Index

A

B

C

D

E

F

G

I

L

M

N

O

P

R

Author's Bio

Jim Gerrish grew up on a grain and alfalfa hay farm in south-central Illinois. He caught grass fever in the mid 1970s and left the grain farming scene, never to return. He is now an independent grazing lands consultant providing service to farmers and ranchers on both private and public lands across the USA.

He received a BS in Agronomy from the University of Illinois and MS in Crop Ecology from University of Kentucky. Jim spent over 22 years conducting beef-forage systems research and outreach while on the faculty of the University of Missouri. With over 20 years of commercial cattle and sheep production on his family farm in northern Missouri, he also has one foot solidly planted in commercial livestock production.

The University of Missouri-Forage Systems Research Center rose to national prominence as a result of his research leadership. His research encompassed many aspects of plant-soil-animal interactions and provided foundation for many of the basic principles of Management-intensive Grazing. Jim was co-founder of the ongoing, very popular 3-day grazing management workshop program at FSRC . These schools have been attended by over 3000 producers and educators from 39 states and 4 Canadian provinces since their inception in 1990. Fifteen other states have conducted grazing workshops based on the Missouri model and Jim has taught in eleven of those states. He typically speaks at 50 to 60 producer-oriented workshops, seminars, and field days around the USA each year.

Together with his wife, Dawn, and four children, he stays in touch with the real world on a 260-acre commercial cow-calf and contract grazing operation. He was deeply involved in the Green Hills Farm Project, a grassroots producer group centered in north-central Missouri and emphasizing sustainability of family farms. His research and outreach efforts have been recognized with awards from the American Forage and Grassland Council, Missouri Forage and Grassland Council, National Center for Appropriate Technology, USDA-NRCS, the Soil and Water Conservation Society, and *Progressive Farmer*.

More from Green Park Press

AL'S OBS, 20 Questions & Their Answers by Allan Nation. By popular demand. Al's Obs' timeless messages presented in question format. 218 pages. $**22.00***

CREATING A FAMILY BUSINESS, From Contemplation to Maturity by Allan Nation. Written with small, family businesses in mind, Nation covers pre-start-up planning, pricing, production, finance and marketing; how to work with your spouse and children, adding employees and partners. For anyone who wants to own their own business. "This is the kind of book I wish I'd had when I started out," Nation explained. 272 pages. **$35.00***

COMEBACK FARMS, Rejuvenating soils, pastures and profits with livestock grazing management by Greg Judy. Follow up to *No Risk Ranching*. Grazing on leased land with cattle, sheep, goats, and pigs. Covers High Density Grazing, fencing gear, systems, grass-genetic cattle, parasite-resistant sheep. 280 pages. **$29.00***

DROUGHT, Managing for it, surviving, & profiting from it by Anibal Pordomingo. Forages and strategies to minimize and survive and profit from drought. 74 pages. **$18.00***

GRASSFED TO FINISH, A production guide to Gourmet Grass-finished Beef by Allan Nation. How to create a year-around forage chain of grasses and legumes A gourmet product can be produced virtually everywhere in North America. 304 pages. **$33.00***

KICK THE HAY HABIT, A practical guide to year-around grazing by Jim Gerrish. How to eliminate hay, he most costly expense in operations - anywhere you live in North America. 224 pages. **$27.00*** or Audio version - 6 CDs with charts & figures. **$43.00**

KNOWLEDGE RICH RANCHING by Allan Nation. In today's market knowledge separates the rich from the rest. It reveals the secrets of high profit grass farms and ranches, and explains family and business structures for today's and future generations. The first to cover business management principles of grass farming and ranching. Anyone who has profit as their goal will benefit from this book. 336 pages. **$32.00***

* All books softcover. Prices do not include shipping & handling

To order call 1-800-748-9808
or visit www.stockmangrassfarmer.com

More from Green Park Press

THE MOVING FEAST, A cultural history of the heritage foods of Southeast Mississippi by Allan Nation. How using the organic techniques from 150 years ago for food crops, trees and livestock can be produced in the South today. 140 pages. **$20.00***

THE USE OF STORED FORAGES WITH STOCKER AND GRASS-FINISHED CATTLE. by Anibal Pordomingo. There are times when supplementing pastures, not replacing them with hy, silage or haylage justifies the beneficial use of stored forages. This is different from cow-calf production. Finishing cattle to the High Select/Low Choice grade on forages alone is not natural. It requires unnaturally good forages and management. This book explains multiple factors to help you determine when and how to feed stored forages. 58 pages. **$18.00***

* All books softcover. Prices do not include shipping & handling

To order call 1-800-748-9808
or visit www.stockmangrassfarmer.com

Name ____
Address ____
City ____
State/Province____________________Zip/Postal Code _______
Phone __________________________

Quantity	Title	Price Each	Sub Total
____	**20 Questions** (weight 1 lb)	**$22.00**	_______
____	**Comeback Farms** (weight 1 lb)	**$29.00**	_______
____	**Creating a Family Business** (weight 1 lb)	**$35.00**	_______
____	**Drought (weight 1/2 lb)**	**$18.00**	_______
____	**Grassfed to Finish** (weight 1 lb)	**$33.00**	_______
____	**Kick the Hay Habit** (weight 1 lb)	**$27.00**	_______
____	**Kick the Hay Habit Audio - 6 CDs**	**$43.00**	_______
____	**Knowledge Rich Ranching** (wt 1½ lb)	**$32.00**	_______
____	**Land, Livestock & Life** (weight 1 lb)	**$25.00**	_______
____	**Management-intensive Grazing** (wt 1 lb)	**$31.00**	_______
____	**Marketing Grassfed Products Profitably** (1½)	**$28.50**	_______
____	**No Risk Ranching** (weight 1 lb)	**$28.00**	_______
____	**Paddock Shift** (weight 1 lb)	**$20.00**	_______
____	**Pa$ture Profit$ with Stocker Cattle** (1 lb)	**$24.95**	_______
____	**Pa$ture Profit abridged Audio -- 6 CDs**	**$40.00**	_______
____	**The Moving Feast** (weight 1 lb)	**$20.00**	_______
____	**The Use of Stored Forages (weight 1/2 lb)**	**$18.00**	_______
____	Free Sample Copy ***Stockman Grass Farmer*** magazine		_______
		Sub Total	_______

Mississippi residents add 7% Sales Tax _______ Postage & handling _______

TOTAL _______

Shipping	Amount
1/2 lb	$3.00
1- 2 lbs	$5.60
2-3 lbs	$7.00
3-4 lb s	$8.00
4-5 lbs	$9.60
5-6 lbs	$11.50
6-8 lbs	$15.25

Canada
1 book $18.00
2 books $25.00
3 to 4 books $30.00

Foreign Postage:
Add 40% of order

We ship 4 lbs per package maximum outside USA.

www.stockmangrassfarmer.com

Please make checks payable to

Stockman Grass Farmer
PO Box 2300
Ridgeland, MS 39158-2300

1-800-748-9808
or 601-853-1861
FAX 601-853-8087